THE PATENT • OFFICE PONY

A History of the Early Patent Office

KENNETH W. DOBYNS

Docent Press

Docent Press
Boston, Massachusetts, USA
www.docentpress.com

Docent Press publishes books in the history of mathematics and computing about interesting people and intriguing ideas. The histories are told at many levels of detail and depth that can be explored at leisure by the general reader.

Cover design by Brenda Riddell, Graphic Details.

Contents

List of Figures

This book honors
all of those employees
of the U.S. Patent Office,
who, over the past
two hundred and some years,
have in their own respective ways
ridden the Patent Office Pony,
and especially
Pasquale J. "Pat" Federico
who would not let us forget.

Acknowledgements

Many people and organizations assisted in the preparation of this book. The vast majority of the information came from the Library of Congress, the National Archives, and the collections of the U.S. Patent and Trademark Office. Copies of many of the illustrations used herein were provided by James Davie of the U.S. Patent and Trademark Office. Additional information has been provided by (in no particular order) the Library of the Historical Society of Washington, DC; the Libraries of the Smithsonian Institution in the National Portrait Gallery and the National Museum of American History; the Washingtoniana Room of the Martin Luther King Library (DC); the Library of the British Patent Office; the Maryland Historical Society Library; the Library of the Museum of the Confederacy, Richmond, VA; the Virginia State Library; the Arlington (VA) Public Library; the New Bern (NC) Public Library; the New York Public Library; the New Orleans Public Library; the Bristol (CT) Public Library; the Pratt Library, Baltimore, MD; the DAR Library; the Scientific American Library; the Court Clerk of Montgomery Co. MD; Brian and Jenny Morse of Medford, OR; the Public Library of Cincinnati and Hamilton County; Cincinnati Historical Society; Archives of the University of Cincinnati; the Constantine (MI) Township Library; Patricia Sluby, Norma Rose, and Ken Dood, all of the U.S. Patent and Trademark Office; Mrs. Noble G. Marshall of Roanoke, VA; the Ellsworth Homestead, Windsor, CT; Donna Siemiatkoski of Windsor, CT; Stephen Loewentheil of Baltimore, MD; Edward G. Fenwick Jr. of Arlington, VA; San Jose (CA) Public Library; Archives of the George Washington University; Archives of Howard University; Clarksville-Montgomery Co (TN) Public Library; the Maine State Library; Library of the American Antiquarian Society, Worcester, MA; John S. Howkins of Baton Rouge, LA; Mrs. William B. Huger of Asheville, NC; and John Smith of no fixed address.

Special thanks are due to my panel of reviewers for their thorough readings of various versions of this book in an attempt to minimize my mistakes. They are, in alphabetical order, Louis Allahut, William T. Bryant, Edward J. Connors Jr., Kenneth W. Hairston, and William Cecil Townsend.

Prologue: The Centennial Celebration

On April 8–10, 1891, the patent system celebrated the beginning of its second century. The celebration was a year late, since the first United States patent statute became law on April 10, 1790. The celebration had been suggested on time, but there was no organization to arrange it. Finally, after the celebration did not occur on time, an ad hoc committee, chaired by patent attorney Robert Washington Fenwick, moved the plans off center, and momentum took over. The Congress of Inventors and Manufacturers of Inventions to Celebrate the Beginning of the Second Century of the American Patent System had its first session on Wednesday afternoon, April 8, 1891, at the Lincoln Music Hall, Washington, D.C. The first session was presided over by Benjamin Harrison, President of the United States.[1]

Charles E. Mitchell

Charles Elliott Mitchell, Commissioner of Patents, was a featured speaker at the first session,[2] speaking on "The Birth and Growth of the American Patent System." He noted that the Constitutional Convention, late in its session, decreed that the exclusive rights of inventors could be protected. He said:

> They thought they were applying finishing strokes and touches to an edifice which was otherwise complete, when they were really at work on its broad foundations. For who is bold enough to say that the Constitution could have overspread a continent if the growth of invention and inventive achievement had not kept pace with territorial expansion. It is invention which has brought the Pacific

[1]See [Har92, p. 23].

[2]See [Mit92].

> Ocean to the Alleghanies. It is invention which, fostered by a single sentence in their immortal work, has made it possible for the flag of one republic to carry more than forty symbolic stars.[Mit92]

He pointed out that under the first patent statute of April 10, 1790, the Secretary of State, the Secretary of War, and the Attorney General were to determine in each case whether a patent should be granted. From April to July they awaited a successful applicant. When he came at last, the three Cabinet officers—Jefferson, Knox and Randolph—sat in solemn dignity to determine that Samuel Hopkins was entitled to a patent for his new method of making potash and pearl ash. Mr. Mitchell maintained that the office being discharged was not unworthy of the tribunal.

Orville Platt

He pointed out that this was the first of 450,000 patents which had accomplished wizard-like transformations in a century, which could not have occurred without the stimulus of patents. When the saddle and pillion are compared with the parlor car, when the tallow candle is compared with the electric light, when the messenger boy is compared with the telegraph and the telephone, it is readily seen that the signing of that first patent to Samuel Hopkins was an act of historic grandeur worthy of the participants.

This was followed by a speech by Senator Orville H. Platt of Connecticut, a champion of the patent system.[3] He said:

> When the patent system was established we were less than four millions of people ... Today we are more than sixty-three millions ... There were no 'mechanics' in the meaning of the word as now used ... Mechanical knowledge was monopolized by the blacksmith, the carpenter, the millwright, and the village tinker ... If you would in a measure form a conception of how large a factor invention has been in this progress, try to imagine what our social, financial, educational and commercial condition would be with an absolute ignorance of how steam and electricity can be used in the daily production of things for our sustenance and comfort; with an absolute ignorance of the steamboat, the railroad, the telegraph, the telephone, the modern printing press, and the machinery in common daily use.[Pla92]

[3]See [Pla92].

Senator Platt pointed out that men have adored God as they dwelt upon the words "And God said, 'Let there be light,' and there was light." But the hours are not all light. Man for centuries spent half his life in night and darkness. It is not irreverent to note that when Thomas Edison said: "Let there be light," man had light anytime he wanted it.

The next speaker, Carroll D. Wright, Commissioner of Labor, pointed out that it required thirty-two days of common labor in 1790 to produce a linen bedsheet. Also that Alexander the Great, after winning the battle of Arabela, could send the news to his capital in the same period of time as could Andrew Jackson after winning the battle of New Orleans. Prior to the day of the telegraph, the speed that man could travel was generally the limit of the speed for transmission of information.[4]

Benjamin Butterworth

Congressman Benjamin Butterworth of Ohio, chairman of the House Committee on Patents, and formerly and subsequently Commissioner of Patents, presented a paper on the effect of the patent system on the growth and development of the United States.[5] He said:

> Our fathers builded even better than they knew. I do not know what they hoped for or anticipated as possible under the [Patent] System, the foundation of which they laid in the Constitution, but this we may believe, that neither the most profound thinker nor the wildest dreamer could have anticipated such marvelous changes and improvements as have been wrought out under our Patent System... If some member of the immortal Convention that framed our Constitution, endowed with the gift of prophecy, had arisen in his place, and in plain speech disclosed what their children would behold at the close of the first century as a result of the power conferred upon Congress in the [patent] clause ... his associates would at once have felt an anxious concern in regard to his mental health ... The wise men in Congress fifty years ago found pleasure in ridiculing and laughing at the 'crank,' Morse, who hung about

[4] See [Wri92].

[5] See [But92a].

> the lobby of the House, insisting that he could use the lightning to transmit messages.[But92a]

Further, said Mr. Butterworth, all of the people of the United States in 1840, with all the means then at their command, could not have harvested one of the present annual corn or wheat crops, and had they succeeded in doing so it would have rotted in the barns for lack of means of transportation to spots where at the same moment famine was reigning. One day's wages of a present-day Boston mechanic would pay the cost of transporting a year's supply of food for his family from Chicago, the great Western market, to Boston. Fifty years ago, one month's salary would not have been sufficient for that purpose.

Mr. Butterworth remarked that we read of the marvelous feats the heroes and demigods of Greek mythology accomplished. If they were true, they would still be eclipsed by the actual possibilities of today. One hundred years earlier, the twelve labors of Hercules had been regarded as something which could be accomplished only by a demigod. The inventor has taught us how to surpass everything that Hercules did. The labors of Hercules, said Mr. Butterworth, would be undertaken by any contractor in good standing in the United States of his day, and he would give bond with approved security to complete the work in half the time required by the son of Jupiter. There is not a blacksmith in the United States, said he, who would consent to use the crude appliances in Vulcan's fabled shop. The fleet Mercury, with his winged sandals, could not keep pace with the messenger of Morse.

Robert Fenwick

Finally, Robert Washington Fenwick presented a paper on the history of the Patent Office.[6] Mr. Fenwick's uncle, Benjamin Fenwick, was the messenger of the Patent Office from 1816 to about 1819, and he was succeeded by Mr. Fenwick's father, Robert Welsh Fenwick. Mr. Fenwick noted:

> It is an interesting fact to relate that in these early days a single pony was kept by the Government for the use of the Patent Office, and that the messenger or clerk rode this pony when he went to the State Department to have the patents signed by the Secretary of State and other officials.[Fen92]

[6] See [Fen92].

The other officials included the Attorney General and the President of the United States. The Fenwick boys would have visited the President to get patents signed.

This single pony for the use of the Patent Office is symbolic of two things. First, the primitive state of the country and of the Patent Office at the time, where the quickest way to deliver messages around the city of Washington was by a boy on a pony. Second, riding the Patent Office pony can be taken as a metaphor for working in the Patent Office, which has been done by many hundreds of people for many generations.

Chapter 1

Protection of Inventors before America was Involved

Hundreds of thousands of years ago, a type of ape arose in Africa which was a little different from other apes. For whatever reason, these apes found that they had much more control over the types of vocal noises they could make than did the other apes of the region. A small group of them invented speech, one of the prerequisites for evolution from apes to humans. The early vocabulary was small, but large enough to make the invention work. We do not know who made this invention, but it had to be a group, because experimental use of speech is necessarily a group project. We know that the group was small, because at that level of development, the apes could not live in large groups. And we know that this group had an advantage in food-gathering, hunting and fighting compared with other groups who could not cooperate so closely as the apes who spoke. It is likely that everyone outside the group who could figure out what was going on and imitate it did so.

Tens of thousands of years ago, men invented a new tool-making process when they discovered that stones could be chipped to form new surfaces which made the chipped stones much more useful than the best natural stones which could be chosen from a pile. Knives made from these chipped stones could be used to butcher animals and to carve pieces of wood into much more useful shapes than could be obtained by breaking and burning. In each tribe, a few members were better flint-chippers than the rest. They became artisans for the tribe. This may have been the first invention to originate a manufacturing industry, small and scattered though it was.

Mere thousands of years ago, men discovered that food-gathering was easier if the seeds of food plants were gathered and planted in one small area, and if wild animals, which had previously been hunted for meat, were gathered into herds and tended until the time when they were to be slaughtered for meat. Agriculture had been invented.[1]

Cuneiform Writing

Writing was invented, and mankind passed from the era of prehistory to the era of history by the availability of the written record. Evidence suggests that writing was invented for use on receipts for grain at temple granaries. At first, perhaps one grain of barley would be withdrawn from each bushel furnished to the granary and wrapped in a clay wrapper, which was then impressed, while wet, with the seal of the temple. When the dried clay envelope was offered to prove the amount of grain which had been furnished, the wrapper had to be broken open to check the amount and type of grain actually furnished. Some bright soul decided that life would be simpler if the wet clay were marked with symbols indicating "3 barley, 4 wheat" at the time the envelope was created, so that it would not be necessary to break the dried envelope to check the contents. But the real genius was the one who decided that, with the symbols on the outside of the sealed clay envelope, it was unnecessary to have actual grain inside. Symbols could be embossed on any clay tablet to indicate the amount of grain furnished, or any other expression for which a symbol was chosen.[2]

The human race progressed slowly, profiting by the useful inventions of individuals. The race is incapable of making inventions. Only individual humans can make inventions. Unless an individual can profit from making an invention, his incentive for working to perfect the invention is greatly reduced. Some inventions give their inventor a competitive advantage in his trade, usually provided that he does not reveal these inventions to his competitors. Such inventions are subject to being lost when the inventor dies, and humanity does not progress on their account.

[1] See [DC63, Chapter 1].

[2] See [Edi80, p. 90] and [Edi82, p. 99].

The first hint of a solution to this problem appeared about 500 B.C. in the Greek colony of Sybaris, which became proverbial for its luxury. The Sybarites, who enjoyed living in luxury, made a law that if any confectioner or cook should invent any peculiar and excellent dish, no other artist was allowed to make this dish for one year. He who invented it was entitled to all the profit to be derived from the manufacture of it for that time. This was done in order that others might be induced to labor at excelling in such pursuits. [3]

Greek Cup Bearer

But most of the ancients did not get the point. In Rome in about 30 A.D., there was a skilled artisan who made a glass cup that was unbreakable. He was given an audience with the Emperor Tiberius Caesar (A.D. 14–37) to show his invention. The artisan asked Tiberius to hand it back to him, and then threw it on the floor, which was paved with stone. Tiberius was astonished. But when the artisan picked up his cup from the ground, it was merely dented like a bronze bowl. The artisan took out a little hammer and made the cup quite sound again without any trouble. After doing this the artisan thought he was destined for greatness, especially when Tiberius asked: "Does anyone else know how to make glass like this?" But when he answered that no one else knew how, Tiberius had him beheaded. Why? Because if his invention were generally known we should treat gold like dirt.[4] In one version of the story, the man was also an architect and had aroused the jealousy of Tiberius, who saw to it that his name was stricken from all public records.

Filippo Brunelleschi

The great architect Filippo Brunelleschi began in 1419 to build the massive dome of the cathedral in Florence, as well as other contemporary architec-

[3] See [Edi80, p. 96] and [Edi82, p. 99].

[4] See [DC24, Book 57] and [Pet69, Book 36, Chapter 66].

tural projects. The architect had found some difficulty in bringing the heavy stones over the River Arno into Florence for his work. He invented a new kind of boat in which such loads could be effectively hauled over the river. He refused to make his invention public for fear that "the fruits of his genius" be taken by another without his consent.

In 1421, the Gentlemen of the Works requested from the Lords of the Council of Florence an exclusive privilege for Filippo Brunelleschi to make and use his invention on the waters of Florence for three years, following a vote in which 218 black beans were cast in favor of recommending the privilege and seven white beans were cast against it. It is likely that the privilege was granted, making Filippo Brunelleschi the first known inventor to be so rewarded in almost two thousand years.[5] Various Italian city-states granted exclusive privileges to inventors thereafter with some regularity. In 1594, the Doge of Venice granted a twenty-year exclusive privilege to Galileo for his invention of an irrigation machine.

The earliest English patent grant, and probably the first exclusive privilege ever granted to an inventor by the instrument of letters patent, is dated April 3, 1449. It was a license to John of Utynam, who had recently returned to England from Flanders at the command of King Henry VI, to live in England and to work at all arts and sciences without restriction. Because his art of making colored glass had never been practiced in England and because he intended to instruct various subjects of the King in many arts never practiced in the kingdom beside the art of making glass, the King commanded that none of his subjects could use such arts for a term of twenty years without the consent of John. King Henry VI, who was in the process of establishing Eton College and King's College, Cambridge, wanted John of Utynam to create stained-glass windows for these colleges and to teach others how to create them. All these purposes, as well as protection for John, were included in the patent. John of Utynam practiced his art of making colored glass by installing stained-glass windows at Cambridge University and Eton College, none of which survive.[6]

But this 1449 patent was not followed by another English patent for over a hundred years. The first patent in the regular series of English patents was the grant to Henry Smyth on April 26, 1552, for making Normandy glass.[7]

The issuance of letters patent in England was a royal prerogative. The English monarch issued letters patent for many purposes. Originally letters patent were merely open letters from the monarch to anyone who might read them. Privileges were granted by patent when the monarch or the monarch's officers

[5]See [Gay40, pp. 547–9],[Pra46], [PS70], [Fru43], and [Gom46],

[6]See [Gom46] and [ML11].

[7]See [Gom46].

had been convinced that it would be good to do so. Sometimes this meant good for the country, sometimes it meant good for the monarch, and sometimes it meant good for the officer. But they were always privileges granted by the monarch, not rights due to the recipient.

During the reign of Queen Elizabeth I, the royal officers were generally diligent in their attention to the industrial and economic aspects of English life. Elizabeth issued patents for inventions which created beneficial results and technological progress for all England, but she also issued them for things which were already in common use in well established industries, giving some favored courtier or large contributor to the royal purse the exclusive privilege of selling salt, iron, paper, currants, sulfur, etc. This led in 1601, near the end of her life, to heated protests in the House of Commons. She gave a magnificent speech to Parliament in which she expressed her ignorance that her subjects had been oppressed by her grants, and she promised reform. She promised that the validity of her patents henceforth could be tested in the common-law courts, which previously had not been possible.[8]

Queen Elizabeth I

This speech cut off the serious protests in Parliament for a while. However, the following year, the celebrated case of Darcy v. Allin[9] was brought in the common law courts, for infringement of a patent for playing cards, which were a common article of commerce before the patent was issued. The verdict found that the patent was bad in law.

In 1624, Parliament passed the Statute of Monopolies,[10] which made void all future grants of monopolies, but expressly excepted grants for 14 years or less for the sole working and making of any new manufacture within the realm. This statute became the basis for all subsequent patent grants in countries whose law is derived from English common law, whereby patents are distinguished from illegal monopolies. However, while the Statute of Monopolies limited the kinds of monopolies that the monarch could grant, it did nothing to give an inventor a right to a patent. An inventor still could get a patent only on the whim of the monarch.

After praying for the royal boon of a patent, and paying a very large sum of money, the inventor might get a patent. This gave him merely the right to

[8]See [Gom46], [Hum78, v. 4], and [Mac56, v. 1].

[9]See Darcy v. Allin (or Allein), (1599) 74 *ER* 1131, (1602) 77 *Eng Rep* 1260 and (1599).

[10]See [Jam24, Statute 21, Chapter 3].

go to court to try to prove the validity of his patent before hostile courts. This system worked so poorly that during the 150 years following the beginning of the numbering of English patents in 1617 scarcely more than 1000 patents were issued in England.

Peter Chamberlin, M.D.

In England, in 1598, physicians and surgeons were in two distinct professions. The profession of surgeon was an outgrowth of the profession of barber, resulting in the modern red-and-white barber pole, reminiscent of bloody bandages placed around a white pole to dry. The physicians of the day would not perform surgery. It was 1598 when Peter Chamberlen the Elder was inducted into the guild of barber surgeons. It is known that the first practical obstetrical forceps were invented by a member of the Chamberlen family, and the probable inventor was Peter Chamberlen the Elder. Although he was a barber surgeon rather than a medical doctor, he was surgeon to Queen Anne, wife of King James I of England. He had such an obstetrical reputation that he aroused the envy of the Royal College of Physicians, who had him thrown into Newgate Prison for practice of medicine without a license. He was released from prison by order of his patroness the Queen. He maintained his obstetrical forceps as a trade secret and was able to deliver the babies of women who would otherwise die without the forceps. He died in 1631. Peter Chamberlen the Elder was succeeded in the family secret by his brother Peter Chamberlen the Younger, another barber surgeon with a fine obstetrical reputation, who was constantly embroiled in controversy with the Royal College of Physicians. Peter Chamberlen the Younger had a son, also named Peter Chamberlen, who was the first of the family to obtain a medical degree (actually several medical degrees), and was thus the first Doctor Peter Chamberlen. In 1628, he became a member of the Royal College of Physicians. He had a large obstetrical practice. It was believed at the time that he had a special instrument with which he assisted women in labor. Doctor Peter Chamberlen was a shrewd businessman. An investigation made at the time revealed that he frequently demanded a large fee for his services. Three sons of Doctor Peter Chamberlen were obstetricians—Hugh Senior, Paul, and John Chamberlen. They continued the use of the trade secret forceps, giving the family a continued monopoly in the safe delivery of women in labor. In

1670, Doctor Hugh Chamberlen Sr. offered to sell the secret to the personal physician of the king of France, but the sale was not made. He did later sell the secret to a Dutch physician, who maintained an obstetrical monopoly in Amsterdam for another sixty years, until another Dutch physician published the secret of the forceps in 1732. For over a century, women around the world had died in childbirth because the Chamberlen family had no effective way of profiting from the family invention except by maintaining it as a trade secret.[11]

In the England of 1643, one Joseph Jenks (sometimes Jenckes), iron founder, had great inventive skills and little chance to profit from them.

[11]See [Ave82] and [Dil53].

Chapter 2

Invention Comes to British Colonial America

Joseph Jenks Sr., age 41, iron founder, of Colebrook, Buckinghamshire, widower with two sons, was persuaded in 1643 to emigrate from England to Lynn, Massachusetts, to operate an iron-smelting and foundry business. In that year, Robert Bridges had taken bog iron ore found in Saugus, Massachusetts, to London and persuaded a group of wealthy English gentlemen and merchants to join him in forming the Company of Undertakers for the Iron Works. The company advanced £1,000 to commence the work. The Company of Undertakers chose Mr. Jenks to go out from England and operate the iron works. Mr. Jenks left his two sons, Joseph Jr. and George Jenks, in England with instructions to join him in America later. He successfully set up the foundry, and personally cast the first article, an iron pot holding about one quart. This pot survives in the Lynn Public Library, Lynn, Massachusetts. The iron works apparently operated for about twenty-five years before it became unprofitable.[1]

In 1641, the Massachusetts Bay Colony granted an exclusive right for the manufacture of salt. It appears that this was for the establishment of an imported industry in Massachusetts, and not for an invention.

But in 1646, Massachusetts granted its first exclusive right for use of an invention. The inventor was Joseph Jenks Sr. The General Court recognized that he had made speedier engines for water-mills and also mills for making scythes and other edged tools, and it allowed him fourteen years without disturbance from others who might set up similar inventions. Mr. Jenks purchased the

[1] See [JEJ34], [SS43], National Cyclopedia of American Biography 22:58-59, and [Edi91b, p. 243–45].

right to manufacture scythes at the iron works, and in 1655 he was granted a second exclusive right for seven years to manufacture an improved grass scythe. Apparently, the common scythe of the day was short, thick, heavy and slow. Mr. Jenks made a scythe blade which was thinner and longer and was thickened on the back side for support. For over 300 years the scythe of commerce remained substantially unchanged in shape from that of Mr. Jenks.

Pine Tree Shilling

In 1652, Massachusetts was short of coinage for use in its internal commerce. It decided to coin its own money, despite the fact that the English policy, at least unofficially, prohibited the colonies from coining their own money. Joseph Jenks Sr. was chosen to make the dies for striking the coins. He made dies for threepenny pieces, sixpenny pieces and shillings. They were to be of sterling silver, and by weight were to have five-sixths of the silver weight of the corresponding English coins. This lesser weight would tend to prevent their export from the colony for their silver value. Each was stamped with "Massachusetts" and a pine tree on one side, and on the other side "New England, Anno 1652," together with the number of pence in Roman numerals. There is a story that Sir Thomas Temple, representing the interests of the Massachusetts Bay Colony, showed samples of the coins to King Charles II. When the King asked what kind of tree was represented on the coins, Sir Thomas answered that it was a royal oak tree, the tree which saved the King's life. The King answered that the colonists were "a set of honest dogs," and proceeded with the business at hand.

In 1654, Joseph Jenks Sr. built a fire-engine for the city of Boston to deliver water in case of fire. There were few such engines in the world, and Paris did not get its first for another 50 years.

Joseph Jenks Sr., died in his early eighties, leaving a large family of descendants. His son Joseph Jenks Jr. came over from England in 1645, two years after his father, to operate iron forges and saw mills. He could not find sufficient water power to operate his mills available in Warwick, Rhode Island, so he moved in 1671 to the vicinity of Pawtucket Falls in Rhode Island to build a mill, and incidentally to found the town of Pawtucket, Rhode Island. Joseph Jenks III was a distinguished citizen of Rhode Island. His two principal claims to fame must have been that he was the royal governor of Rhode Island from 1727 to 1732, and that he stood seven feet two inches in his stocking feet. We shall hear more of this family later.

Thomas Masters, a Pennsylvania Quaker, and his wife Sybilla applied for a patent in 1715, not for the colony of Pennsylvania, but back in the home country. He obtained British patent number 401 of 1715, actually secured in 1716, for a donkey-driven and a water-powered engine for shelling and pulverizing maize into meal. There are two remarkable facts about this patent. The first is that it may have been the first British patent granted to a resident of the American colonies. The second is that the invention was made and the patent was prosecuted in England not by Thomas Masters, but by Sybilla. This may have been the first patent ever granted anywhere for an invention which was made by a woman. They recorded this patent in Pennsylvania in 1717,[2] and sold the corn meal, called "Tuscarora Rice," as a cure for consumption.[3]

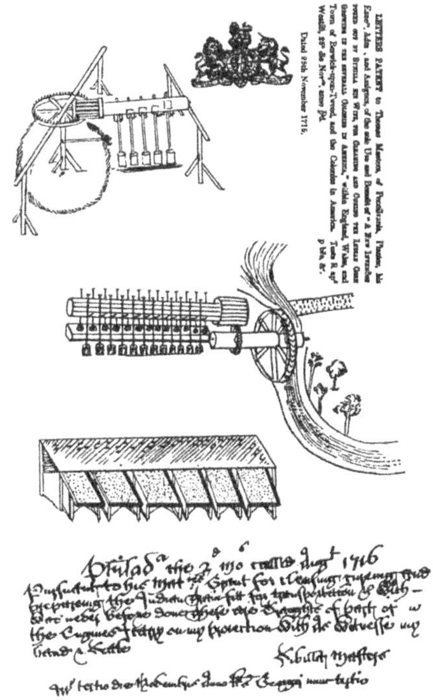

Patent Drawings for Thomas Masters' Engine

In Connecticut in the mid-1700s, there lived a superb craftsman named Abel Buell, equally skilled as a jeweler, type founder, and counterfeiter. He had been convicted of altering the bills of public credit of the colony and was sentenced to life imprisonment. When subsequently paroled to dwell with his family in Killingworth, he applied his skills to the legitimate crafts in his repertoire. He applied to the Connecticut legislature in 1766 for a special premium, which they had promised for useful discoveries, noting that he had discovered a method of grinding and polishing crystals which were found in the colony. He requested, as his special premium, that he be restored to the liberties which he had justly forfeited to enable him to carry on the business involved in working with such crystals. He was restored to his liberties, conditional upon giving bond. Then, in 1769, he petitioned the legislature again, noting that he had developed the art of type founding,

[2]See [Edi97, pp. 370–71], [otCoP49] and [Gri91, pp. 71–72].

[3]See [Tol71] and [Nee84].

which was known to few people, even in Europe. He wanted to set up a type foundry, and asked, as his special premium, a lottery to provide him with funds to set up the type foundry. Specimens of his type are still in existence. The legislature advanced him £200 to set up his foundry, on condition that he not depart the colony for seven years. By 1777, he had fled Connecticut because of debt, indicating that the type-founding business was not a total success. He appeared again in the records of Connecticut as a resident of New Haven in 1785 and continued there as jeweler, instrument maker, draftsman, silversmith and machinist until his death in 1822.[4]

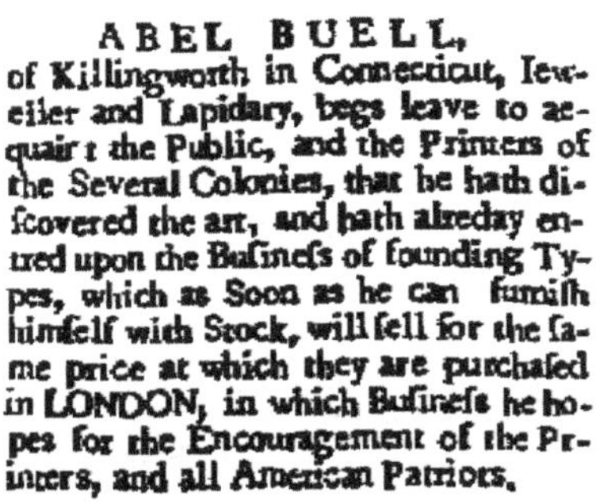

ABEL BUELL,
of Killingworth in Connecticut, Ieweller and Lapidary, begs leave to acquaint the Public, and the Printers of the Several Colonies, that he hath diſcovered the art, and hath alreday entred upon the Buſineſs of founding Types, which as Soon as he can furniſh himſelf with Stock, will ſell for the ſame price at which they are purchaſed in LONDON, in which Buſineſs he hopes for the Encouragement of the Printers, and all American Patriots.

Abel Buell's Type Font

There seems to have been a recognition of the importance of invention in much of British colonial America which was generally absent in French and Spanish colonial America. Most of the early legislation to reward inventors was in New England, particularly in the colony of Connecticut. There were numerous other monopolies granted in colonial times, in such areas as working iron and steel, raising and using silk, and making paper, which are mentioned in the records, without sufficient information to make them broadly interesting.

The vigorous American interest in invention seems to have blossomed with the serious desire for and attainment of independence, beginning about 1775.

[4]See [otCoP50], [Moo55], and [Wro26].

Chapter 3

Independent America Explores Invention

When the winds of revolution began blowing through the British colonies, the inventors and manufacturers took up the challenge of preparing the infant nation for the inevitable fight.

In 1762, the Connecticut legislature had apparently granted money to continue the mining of iron ore from the iron beds at Salisbury, which they had sponsored in some fashion for many years. The Council of Safety in Connecticut decided at the beginning of 1776 to take over the iron furnace at Salisbury for manufacturing iron, casting cannon and cannon balls, etc. They continued to finance this activity during the years of the American Revolution, exempting important workers there from military service.[1]

Understanding the necessity of thwarting the British fleet, David Bushnell of Connecticut devoted his efforts to the development of a submarine, which was styled Bushnell's Turtle. The submarine was to be used to attach explosive canisters, or torpedoes, to the hulls of enemy ships. Hand-driven screws and a single rudder would allow the lone op-

Bushnell's Turtle

[1]See [otCoP50].

erator to approach a ship while underwater and to attach the canister to the hull by driving a heavy screw into the hull, to which the canister torpedo was attached. A timer-operated fuze would set off the explosive many minutes or hours later, allowing the submarine time to depart safely. Mr. Bushnell appeared before the Council of Safety in Connecticut in February 1776 to ask its support and potential reward in case of success. The Council ordered that the Treasurer pay £60 to Mr. Bushnell to improve his invention. Subsequently, Mr. Bushnell made several attempts to blow up British ships in New York harbor. However he failed because the ships were sheathed in copper, and he could not drive the canister screw into the hull through the copper.[2] How quickly might the Revolution have been won if it had worked!

One cannot run a war or a government without paper, and the supply from England was soon cut off. The Continental Congress quickly passed a resolution that the paper-makers of Pennsylvania would not be allowed to join the military battalions, their services being needed to make paper for the revolution. But while there were a number of paper-makers available, there was apparently only one person in Pennsylvania who knew how to make the wire-screen molds needed by the paper-makers for turning their pulp into paper. That person was Nathan Sellers of Philadelphia. He was ordered by the Continental Congress to be removed from Colonel Paschall's Battalion in New Jersey and to be returned to Philadelphia to make molds for the paper makers.[3]. The Sellers family is another family that we will hear from later.

Nathan Sellers

In these early days, it was the practice to card wool by hand in order to align the fibers for subsequent spinning. Two hand-held brushes or cards were used, each tediously handmade by inserting wire teeth or wire bristles into a leather backing. Every household with a spinning wheel needed its own pair of wool cards. Oliver Evans, of Newcastle, Delaware, began during the American Revolution to develop a self-feeding machine to punch holes in the leather and insert and fix the teeth in the card, making a one-thousand-tooth card each minute. In 1786, he petitioned the Maryland legislature for a 25-year monopoly for making card machines,

[2]See [otCoP50], [Abb81], and [Weg63].
[3]See [For48, col. 1619].

and for making machines to automate the operation of flour mills, a little invention he was to spend much of his life trying to profit from.[4]

And in an early preliminary shot in the Rumsey-Fitch steamboat fights, James Rumsey, of Berkeley County, Virginia, filed a petition with the Maryland legislature in November 1783 requesting a monopoly for an unstated term of years in his various inventions.[5] One of these invention was "for propelling boats on water, by the power of steam, which has already been accomplished in experiments on a boat of about six tons." This became confused with his request in January 1785 to the same legislature for a ten-year monopoly for "navigating boats against the current of rapid rivers," which appears to have referred to a system for using water power to drive poles to pole a boat up a rapid river. In September 1784, George Washington provided Mr. Rumsey with a letter,[6] much cited later, in which he stated that he was an eyewitness to an actual experiment in which Mr. Rumsey's boats worked against the stream, by mechanism and small manual assistance, against rapid currents. Whether or not any steam power was used in the experiment witnessed by Washington, he made no mention of steam in his letter.

According to Mr. Rumsey's chronology,[7] much disputed by the Fitch interests, Rumsey proposed a steamboat in the summer of 1783, and in December 1785 prepared a steam engine and boat to test. He was prevented by the icing up of the Potomac River in January 1786 from testing it then. Over the winter, he improved his equipment. He had parts made for the boat in February 1786. The boat was tested in March 1786 and performed, but quite imperfectly. Mr. Rumsey was fortunate to attract the attention and support of many of the important men of the time, the most prominent of whom was Benjamin Franklin, who set up the Rumsean Society to support him. Following Mr. Fitch's chronology,[8] equally disputed by the Rumsean interests, Fitch conceived his invention in 1785 and exhibited a model of his steamboat to the American Philosophical Society in September 1785. Fitch published a descrip-

James Rumsey

[4]See [otCoP49].
[5]See [otCoP49].
[6]See [otCoP49].
[7]See [otCoP49] and [otCoP50], and [How45].
[8]See [otCoP49],[otCoP50], [Fit76], [Tho14], and [How45].

tion and illustration of his steamboat in the Columbian Magazine in December 1786.

Different students of the Rumsey-Fitch controversy can and have reached opposite conclusions as to who should be given priority. Neither was the first to propose a steamboat. Among the numerous prior proposers was Thomas Paine, the Revolutionary War pamphlet author, who proposed construction of steamers to the Continental Congress in 1778.[9] Nor was either the first to build and operate a steamboat. A steamboat was constructed and operated under the direction of the Marquis Jouffroy in the river Soane in France in 1781.[10] What is important is that Rumsey and Fitch fought for monopolies on their inventions to be granted by at least six state legislatures in the days before the beginning of the national patent system. The legislatures had no rules to govern their granting of monopolies. A monopoly could be granted to encourage the development within the state of technology which was well known abroad, even though the person granted the monopoly was admittedly not the first inventor. Or a monopoly could be granted to an inventor who alleged that he was the first, without adequate means to check the accuracy of the allegation. And after long and involved efforts to lobby a state legislature to grant a monopoly, the monopoly granted was not valid outside the state. It could be written out of the law if another petitioner could convince the legislature that he had a better case or more influential friends.

In a country of only four million people, there was not a market large enough to support extensive manufacturing unless the market could be as wide as the nation. To support this, inventors needed a monopoly as wide as the nation. It was necessary, but perhaps not a clearly seen necessity, to provide patents as wide as the nation to support markets as large as the nation.

[9] See [otCoP49] and [Moo55, p. 318].

[10] See [Moo55, p. 318].

Chapter 4

To Promote the Progress of the Useful Arts

In 1787, the loose confederation of thirteen independent states had proven itself an inadequate government for a thinly populated nation whose population mostly hugged the eastern coastal areas of the infant nation. Such roads as existed were long narrow gullies of mud when wet and long narrow ruts of dust when dry. The preferred method of travel was still by water where possible.

John Fitch, farmer, surveyor, clockmaker, silversmith, gunsmith, frontiersman, mapmaker, was hit with the idea of a steamboat in 1785. With the monomania which sometimes affects inventors, he devoted most of the rest of his life to promoting his invention. By 1787, he had built and was ready to demonstrate his steamboat in Philadelphia, to enable easier travel by water.

John Fitch

On May 14, 1787, delegates from the thirteen states met in Philadelphia in the Constitutional Convention, intending to draft a constitution to replace the nearly inoperative Articles of Confederation. George Washington presided at the convention, after spending four days in traveling the 150 miles between Mount Vernon and Philadelphia. On May 29, 1787, Edmund Randolph of Virginia opened the business of the convention by submitting a series of resolutions known as the "Virginia Plan." Then Charles C. Pinckney of South Carolina presented the convention with a proposed plan for the federal government which he had prepared. There was no mention in either of these plans of any power to grant patents. The plans were referred to a committee, and the committee subsequently reported in favor of Mr. Randolph's plan. However, the Randolph plan was amended in the committee of the whole house. No reference to a patent power was made yet. Discussion of the "Virginia Plan" was postponed until William Patterson of New Jersey could submit a plan. Both of these plans were referred to the committee of the whole, which reported again in favor of Mr. Randolph's plan as the basis of the Constitution. After the report was debated for over a month, all of the proceedings of the convention up to that time were referred to a committee of detail appointed for the purpose. Thirteen days later, the committee made a report, but still there was no provision for granting patents. Practically the entire Constitution had been thoroughly considered before any member of the convention suggested the power of granting patents.

James Madison

On August 18, 1787, James Madison of Virginia arose in his place, and submitted, for reference to the committee of detail, additional powers to be added to those previously proposed for the legislature. Among these powers were "to secure to literary authors their copyrights for a limited time," and to "encourage by premiums and provisions, the advancement of useful knowledge and discoveries." On the same day Charles Pinckney of South Carolina also submitted a number of propositions, among which were: "to grant patents for useful inventions," and "to secure to authors exclusive rights for a certain time." It is noteworthy that South Carolina was the only state which had passed general legislation allowing the grant of patents without special act of the state legislature. The propositions of both these gentlemen were soon referred to the committee of detail.[1]

Mr. Fitch invited Dr. William Samuel Johnson, a member of the Constitutional Convention from Connecticut, to ride aboard his steamboat in the Delaware River at Philadelphia on Wednesday afternoon, August 22, 1787. Dr. Johnson invited others, including Rufus King of Massachusetts, to accompany him. On the day of the demonstration ride, the convention adjourned to allow the members to see the show. The boat operated under its own steam power,

[1]See [Cam91b, p. 241] and [Fen29].

and several of the members rode on it. Fitch noted in his autobiography that few of the members of the convention failed to call and see his steamboat.[2] The only one he was sure did not show up was General Washington. We know that Oliver Ellsworth of Connecticut, later Chief Justice of the United States, was one of the riders.[3]

On August 31, such proposed parts of the Constitution as had not been acted upon were referred to a committee composed of one member from each state. Among these undisposed parts were the propositions to give Congress the power to grant patents for inventions. Mr. Madison was on this committee, but Mr. Pinckney was not.

On September 5, 1787, the committee reported and recommended, among other things, that Congress should have the power "to promote the progress of science and useful arts, by securing for limited times to authors and inventors the exclusive right to their respective writings and discoveries." There was no recorded debate on the patent clause. Perhaps, after the demonstration of John Fitch's steamboat, no debate was necessary. The patent clause was agreed to without a dissenting vote. In the final revision of the style and arrangement of the articles in the Constitution, this clause became paragraph 8, section 8, of Article I, where it is now.

Charles C. Pickney

Thus it is seen that the proposals to give this patent power to Congress were made by James Madison and Charles Pinckney. Neither of them seems to have had any special interest in science or the useful arts. Mr. Madison, in one of the Federalist Papers, argued in favor of the patent and copyright clause as follows: "The utility of this power will scarcely be questioned. The copyright of authors has been solemnly adjudged in Great Britain to be a right at common law. The right to useful inventions seems with equal reason to belong to the inventors. The public good fully coincides in both cases with the claims of individuals. The States cannot separately make effectual provision for either of the cases, and most of them have anticipated the decision of this point by laws passed at the instance of Congress."

Time has justified Mr. Madison's Federalist Papers argument.

[2]See [Fit76, p. 178–79].
[3]See [SD01, 8/27/1787].

John Fitch had succeeded in interesting about twenty investors in his plan, and he had persuaded them to take shares in his steamboat company at $50 each. Following his August 1787 demonstration, the company started another larger steamboat with steam-engine-driven oars. John Fitch, who had little real skill as a mechanic, found the skills he needed in Henry Voigt, a Dutch watchmaker of Philadelphia. On the first trial of this boat on the Delaware in the summer of 1788, the boiler began to leak and the engine stopped when John Fitch and Henry Voigt were in sight of their destination. The next day Fitch and Voigt brought the disabled boat back to Philadelphia on the tide. Naturally, the other boatmen on the river ridiculed their boat. But Fitch and Voigt continued their efforts, and in October they took a number of passengers on a pleasure trip to Burlington, New Jersey, at a speed of about four miles an hour.[4]

John Fitch's Steamboat

At this time in America, there were few skilled mechanics. A young English medical doctor, William Thornton, was one of the steamboat company. He stated years later that there was not a single engineer in the company, and that before the work started he was the only one who had even seen a steam engine. That was as a boy in England.[5] They were forced to make engineers from common blacksmiths. A small group of members of the company undertook to increase the speed of the boat from three or four miles an hour to eight miles an hour, at their own expense, in exchange for half the shares of the remaining members if they were successful. They worked at this for about 12 months.

[4]See [Fit76], [Tho14, *Short Account*], [SS43], and[BDW88].

[5]See manuscript papers of William Thornton, Library of Congress, Washington, D.C. sheets 505–508, William Thornton draft letter 1/1/1810.

On April 16, 1790, a test was arranged. A mile was measured off on Front street, sometimes called Water street, and the boat was found to go at the rate of eight miles an hour. It afterwards went 80 miles a day. The Governor and Council of Pennsylvania were so much gratified with the experiment that they presented them with a superb silk flag. Fitch said that he and Voigt thought themselves Lord High Admirals of the Delaware that day. About this time, the company sent Mr. Fitch to France, at the request of Aaron Vail, then American consul at L'Orient, who was one of the company and who wished to introduce the invention into France. Because France was in the midst of revolution when Fitch arrived, no men could be found to build boats. Therefore, Mr. Fitch departed France. Fitch later made further alterations to his boat, and it performed tolerably well, but it required still further alterations. Mr. Fitch, however, was not able to obtain the necessary financing to perfect his invention, and the project faltered.

John Fitch's belief in the power of steam remained firm. In June 1792, he wrote to David Rittenhouse, a member of the company. He said that steam power will be the mode of crossing the Atlantic in time, whether he perfected it or not. He suggested that Mr. Rittenhouse buy his lands in Kentucky to enable him to complete the great undertaking.[6]

When John Fitch called upon a blacksmith who had worked on his boat to discuss their work, he concluded by saying that although he might not live long enough, some of the onlookers would see the time when steamboats will be the preferred mode of passenger conveyance, and would be particularly useful in travel on the Mississippi. When he left, an onlooker suggested that it was a pity that Fitch was crazy.

John Fitch petitioned Congress for a patent for years, even before the Constitution was signed. He was to await a patent statute before this desire could be satisfied.

[6]See [Edi46, p. 1].

Chapter 5

The First U.S. Patent Statute

John Fitch was not alone in petitioning Congress for a patent. The Constitution said that Congress had the power to grant patents and copyrights, and there were inventors and authors out there ready to ask. Numerous petitions were filed with the first session of the First Congress in 1789, requesting patents and copyrights. Apparently, none was granted. In the first session, a patent bill of some type was proposed in the House, but it was not passed, and almost all record of it was lost when the British burned some of the Congressional records in 1814.

On January 8, 1790,[1] President Washington, in his State of the Union message to the second session of the First Congress, meeting in New York City, recommended giving encouragement to the introduction of new and useful inventions from abroad, as well as to the exertion of skill and genius in producing them at home. Three days later, the Senate replied that the introduction of inventions from abroad and the exertions of skill and genius in producing them at home would receive early attention. The next day, the House replied that the promotion of science and literature would contribute to the security of a free government, and that it would not lose sight of such worthy objects.

On January 25, 1790, the House appointed a committee to draft a patent statute. The committee consisted of Aedanus Burke of South Carolina, Benjamin Huntington of Connecticut, and Lambert Cadwalader of New Jersey. On February 17, Mr. Burke reported a bill to promote the progress of the useful arts, and it was given its first reading.

[1]See [Fed32] and [Gal34]

Aedanus Burke

Meanwhile, Francis Bailey petitioned Congress for a patent for certain punches which were to prevent counterfeiting, and his petition was supported by Alexander Hamilton, Secretary of the Treasury. The House decided to delay consideration of Mr. Bailey's request until a general patent statute could be passed.

On March 4, the House appointed a Committee of the Whole House to consider the proposed patent law. The committee decided not to allow appeal to a jury from a decision of appointed referees on interference issues, as had been proposed. The House gave the bill its third reading and passed it on March 10, and sent it to the Senate the next day. The Senate passed the bill with twelve amendments on March 30, and sent it back to the House on the same day. On April 3, the House agreed to all but one of the twelve amendments and notified the Senate on April 5. On April 5, the Senate agreed with the House version. On April 10, 1790, the first patent statute was signed into law by President Washington.

Benjamin Huntington

It has always been known by those associated with the Patent Office that it is an old-line agency, with origins dating back to the earliest days of the nation. We have not always realized how true this is. The first patent statute was passed on April 5, 1790, by the Congress of the twelve United States and signed into law on April 10 by the President. Rhode Island ratified the Constitution and joined the Union as the thirteenth state on May 29, 1790, 49 days after the first patent statute was in effect. We had our first patent law before we had our thirteenth state.

The first patent statute did not create a Patent Office. It directed applicants to file a petition for patent with the Secretary of State. Then the Secretary of State, the Secretary of War, and the Attorney General were to determine if they, or any two of them, thought "the invention or discovery sufficiently useful and important" to merit a patent.[2] The applicant for patent was required to provide a specification containing a written description of the invention, together with drafts (presumably technical drawings) and an exact model of the invention, if the nature of the invention would allow a model. The description and model had to be good enough to enable a skilled workman to make and use the invention, so that the public could have full benefit of the invention after the expiration of the patent.

When the designated committee had decided that a patent was appropriate and they had caused the legal instrument styled letters patent to be made out, the instrument was referred to the Attorney General for his review as to legal sufficiency. After his approval, the patent was sent to the President of the United States for his personal signature, and then back to the Secretary of State for his personal signature attesting that he had caused the Great Seal of the United States to be affixed. The patent was then delivered to the patentee, and it ran for 14 years. Early patents are much desired by latter-day autograph hunters for their collections of valuable signatures.

The cost for all of this was surprisingly low. There was a charge of 50 cents for receiving and filing the petition, and a charge of 10 cents per 100 words for copying the specification to copy sheets. The cost of making out the patent was two dollars, with an additional cost of one dollar for affixing the Great Seal, and finally 20 cents for intermediate services leading to the endorsement of the patent with the date of delivery to the patentee. All of this money, usually amounting to between four and five dollars, was paid not to the government, but to the government employees who provided the services. In effect, this meant payment to the Chief Clerk of the Department of State.

Lambert Cadwalader

The officers given the duty of determining if an invention was sufficiently useful and important to merit granting a patent were Secretary of State Thomas Jefferson, Secretary of War Henry Knox, and Attorney General Edmund Randolph. They called themselves collectively the "Commissioners for the Promotion of Useful Arts," or the "Board of Arts," although others called them the Patent Commission or Patent Board.[3]

[2]Act of April 10, 1790. *Statutes at Large* 109

[3]See [Fed36].

We do not know how long it took for proper patent applications to be filed with the Patent Board. Many petitions had been filed with Congress, but there were additional requirements, such as models, and possibly a more complete description of the invention, beyond those of a mere petition to Congress. The first United States patent was issued on July 31, 1790, but it may not have been based on the first application which was filed. As will be noted many times between here and the chapter which covers it, the records in the Patent Office were completely destroyed in a fire on December 15, 1836. All that we know about this period comes from scattered records preserved in other locations.

Only three patents were granted in 1790. The slow beginning may have had more than one cause. Certainly, the Patent Board was severe in the application of the law to applications. Thomas Jefferson observed,[4] a few months after the 1790 law came into effect, that the law had given a spring to invention beyond his conception. Many of these inventions he found trifling, but some he found of great consequence. There could also be other reasons. The members of the Patent Board were busy men, with many other obligations. And there were no American rules as to what should be patentable and what should not. The Patent Board had to invent a patent system and a set of rules of patent practice before they could truly begin investigating patent applications. The patent system they invented at the outset was not perfect, but it may have been the best ever seen at that time.

Edmund Randloph

Henry Knox

Thomas Jefferson

Commissioners for the Promotion of Useful Arts

In July 1790, the Patent Board approved a patent for Samuel Hopkins (1743–1813), a pot ash maker of Philadelphia, Pennsylvania. On July 31,

[4]Thomas Jefferson to Benjamin Vaughn, 6/27/1790. See [Lip03, v.8] and [BG90, v. 16]

1790, at the first national capital in New York City, the papers were signed by all the necessary parties and issued as a patent for a method of making potash and pearl ash by burning wood ashes a second time before dissolving them to extract potash.[5] Because the invention was in chemistry, probably no model was furnished. With no model required, it would have been possible to file the Hopkins application more quickly than was possible with applications for mechanical inventions. Hopkins offered a five years license to use his invention for $200, payable as two tons of potash. His degree of licensing success is unknown.

Only a week later, on August 6, 1790, also in New York City, a second patent was issued, to Joseph S. Sampson, for manufacturing candles. Like many patents to be issued over the next two centuries, the second patent disappeared into oblivion. If anyone other than the inventor operated under the second patent, that fact is lost in the mists of time. But near the end of the year, from the new national capital in Philadelphia, the third and last patent of the year was issued on December 18, 1790, to Oliver Evans, an inventive genius of the first magnitude. There is more to be said about him in following chapters.

In a letter in later years,[6] Thomas Jefferson wrote that as a member of the Patent Board for several years, he had observed with what slow progress a system of general rules could be matured. However, the board did manage to establish some rules. One was that any known machine should remain available for any use that it might be put to. Another was that a mere change of material of construction of a known device should not entitle one to a patent. A third was that a mere change in form of an old device should not entitle one to a patent. It is interesting to note that modern patent practice allows major exceptions to all three rules, too complex to consider here.

Before the Constitution came into effect, John Jay was appointed Secretary of Foreign Affairs by the Congress, and the Under Secretary was Henry Remsen Jr. (1762–1843).[7] When the State Department was formed under the Constitution, Mr. Remsen soon became its Chief Clerk. Sometime between April 1790, when the first patent statute was passed, and March 1792, when Mr. Remsen left office, he wrote a memorandum on how to conduct patent business. It referred to "patent No. 21" (even though patents were not then officially numbered), and the twenty-first patent was issued in August 1791, making it probable that the memo was written after that date.[8]

---

[5] See [Pay90].

[6] Thomas Jefferson to Isaac McPherson, 8/13/1813. See [Lip03, v. 13].

[7] See [Sch43a], [Ham01], and *N.Y. Evening Post* obituary 2/20/1843.

[8] See [BG90, v. 23].

Henry Remsen, Jr.

He noted in his memorandum that the Patent Board met on the last Saturday of each month. Sampson Crosby, the State Department messenger, was charged with the duty of notifying each of the members about the meeting the day before. At each meeting, the board read each of the applications received since the last meeting, and then the members could consider them for a month. The board would never decide on an application at the end of that month unless all the necessary specifications, drawings and models had been submitted. No models were to be returned to patentees without orders from the Secretary of State. Instructions were provided for proper forms for specifications, and for the proper labeling of models. The method of making entries of dates of receipt, etc., into the minute book was explained.

It was also noted that the petitions received were to be filed in "the desk upstairs in one of the pigeon holes." Another pigeon hole was to be used for the petitions decided upon, and also the drafts of issued patents until they could be recorded. Some of the specifications were said to be in the desk and others were in the closet. Apparently, this desk was the personal property of Henry Remsen Jr., because it remained in his family and was on loan to the Patent Office for an exhibition in 1936. This may be the origin of the tradition that in the early days of the Patent Office, all of the records of the Patent Office were kept in a few pigeon holes of a desk.[9] While not entirely true, it does appear that everything under active consideration would fit into the pigeon holes of Henry Remsen's desk.

State Department Office, 1791, Market & 7th St., Philadelphia, Headquarters of Patent Board

[9] See [Edi85], [Ano86], [Eva19], [Wym15, p. 533], and [Off77].

In July 1791, Thomas Jefferson wrote to General Henry Knox,[10] noting that the members of the Patent Board had agreed that the descriptions for patents should be provided to the members separately at their lodgings, to be examined at their leisure. Thus, reviewing patent applications became an after-hours occupation to be carried out at home. On this occasion, Thomas Jefferson sent General Knox some papers which had already been reviewed by both Jefferson and Randolph. He noted that this was the first set of papers sent around in the fashion agreed to in their meeting, so that a detailed letter of explanation was considered appropriate. Afterwards, no such explanation would have to be provided. The criticisms and amendments made by the members of the Patent Board would be consolidated by Mr. Remsen before the next meeting of the board.

John Stevens

Minutes were kept of the meetings of the Patent Board, but these have disappeared, perhaps burned in 1836. The best notion we have of the total business of the board comes from a memorandum dated March 31, 1792, written by Henry Remsen on his last day in office, [11] in which he gives a detailed account of the business on hand. Although only 47 patents were issued through the end of 1792, at least 114 applications had been filed by the date of the memorandum. This list is incomplete, because it lists only those applications still under consideration, and omits entirely those which had been refused prior to the date of the inventory. A few other interesting facts can be gleaned from this report. John Fitch's patent cost him $4.39, while James Rumsey got six patents at a total cost of $32.18. Samuel Mulliken got four patents at a cost of $16.07. The fees were divided between Henry Remsen Jr., and another clerk, George Taylor Jr., as part of their salaries. George Rogers Clark applied for a patent for propelling boats, but no patent was ever issued. An apparently common fault among applicants noted in the report was the failure to provide proper specifications, drawings, and models.

In November 1790, the Patent Board moved, with the rest of the Government, from New York City to Philadelphia. John Fitch, who spent much of his time in Philadelphia, was there waiting for them, and on November 22, 1790,

[10]Thomas Jefferson to General Knox, 7/22/1791, in JPOS 19:363.

[11]Remsen memo, 3/31/1792, partial copy in National Archives M179 reel 7, full original at Record Group 59 Misc. Corr. (1784–1906) Ltrs. Recd. MARSA-1 Entry 113 Vol. 14, March–April, 1792.

he presented a patent application (or petition) to them. This was not his first patent application. Indeed, he had explained his invention to the Congress under the Articles of Confederation in August 1785, and he petitioned them for an exclusive right to his invention in March 1786. Mr. Rumsey had filed an application for patent with the Patent Board earlier than November 22, 1790.

The Patent Board issued an order on November 23, 1790, [12] that all of the parties who had requested patents founded upon the discovery of applications of steam to useful purposes be given a hearing on the first Monday in February 1791, and ordered notice of the order to be given to John Fitch, James Rumsey, Nathan Read, Isaac Biggs, and John Stevens.

Nathan Read

The board was not satisfied with the precision with which the inventions were described in the applications, and each of the inventors was to transmit in writing to the board a precise statement of his several inventions and the extent thereof. Isaac Biggs alone does not appear in the later list of steam patentees and is apparently otherwise unknown to history.

On April 21, 1791, John Fitch and some of his supporters met the Patent Board at the Office of the Secretary of State at Market and Seventh Streets, Philadelphia, possibly in response to this order. He presented his petition to the board members, and it was read and laid on the table without a single observation on it. [13] Mr. Fitch had learned that the Patent Board had decided to issue patents to all rival applicants, and he asked that the earliest patent might be granted to him, as he could prove that Mr. Rumsey came to his invention 12 months after him. Mr. Randolph said that the earliest applicant must have the earliest patent, thinking, as Mr. Fitch supposed, that Mr. Rumsey had the earliest application since he filed first with the Patent Board. All seemed to Mr. Fitch to assent to Mr. Randolph's assessment. Mr. Fitch believed that the Patent Board was showing partiality toward Mr. Rumsey, and pointed out that he had petitioned to the Congress for an exclusive right in March 1786, long before the Patent Board was even founded. Upon this statement, the members were silent for a time, then Mr. Jefferson said that

[12] See [ES05], quoting extract from Patent Board minutes of November 23, 1790, sent to Evans by Henry Remsen, Jr.

[13] See [Fit76].

they could make no distinction in the patents nor give one preference over another. Mr. Fitch argued against this position, but the Patent Board decided that all patents should be issued on the same day. Fitch noted that he did not like to remind the board that Mr. Randolph had just said that there should be a preference for the earliest application.

Mr. Fitch noted that the judgment of the Patent Board changed as quickly as the weather, and for no other reason than that he showed that he was the first applicant. That, it appeared to him, was a sufficient reason in their judgment why the earliest patent should not go to the earliest applicant. Patents issued on all of the steamboat inventions on August 26, 1791.

After receiving the decision from the Patent Board, Mr. Fitch resolved to leave Philadelphia in late April and to spend no more time or effort to complete his scheme. If several parties had the right to use steamboats, nobody could attract investors to develop them exclusively. By granting patents without determining priority, the Patent Board destroyed the aspirations of all of the inventors, and delayed the potential of American steamboats for years to come.

Mr. Fitch spent some time writing his autobiography and considerably more time drinking and raging at the government. He had an intense mistrust of Thomas Jefferson, and when he donated his manuscript autobiography to a Philadelphia library he insisted that it remain sealed for thirty years, partly to prevent interference with it by Mr. Jefferson's friends.

John Fitch said that it would be possible for 100-ton steamboats to travel the Mississippi under the control of only four men, making the Western Territory much more valuable. Perhaps steamboats would not be developed to this point in his lifetime, but it would be done, and would have been effected by "little Johnny Fitch and Harry Voigt ... even if my country are pleased to see me in rags, penury, and distress."

John Fitch was a tired, disillusioned, bitter, and physically ill man when he went to his land in Kentucky to drink. But even a tired and disillusioned John Fitch could not stifle his inventive mind, and he quietly and privately built a three-foot steam engine, designed to run on tracks—a lost forerunner of the locomotive. Finally, he went to a physician about his physical illness, and was given a dozen opium pills which he could take when needed to relieve his pain. He achieved quick and permanent relief by taking them all at once, [14] probably with a liberal dose of alcohol. He died at Bardstown, Kentucky, in the summer of 1798, before the first steamboat passed the place, but he expressed a wish to be buried there on the banks of the Ohio, so he would

[14]See [SS43] and National Cyclopedia of American Biography 6:63-64.

repose where the song of the boatman would enliven the stillness of his resting place, and the music of the steam engine would soothe his spirit.[15]

James Rumsey was in England to try to exploit his invention there after being given part of the undistinguished lump of patents awarded by the Patent Board. He was on the point of demonstrating his steamboat, when he was invited to speak on December 18, 1792, to a public gathering at Liverpool. The hall was filled to capacity, and he lectured on hydrostatics. Shortly after finishing, while still at the meeting, he suffered a stroke. He died the next day. His steamboat, the Columbia Maid, was set in motion on the Thames soon after his death. Without his support, and with his financial affairs thrown into chaos by his sudden death, nothing came of the invention.[16]

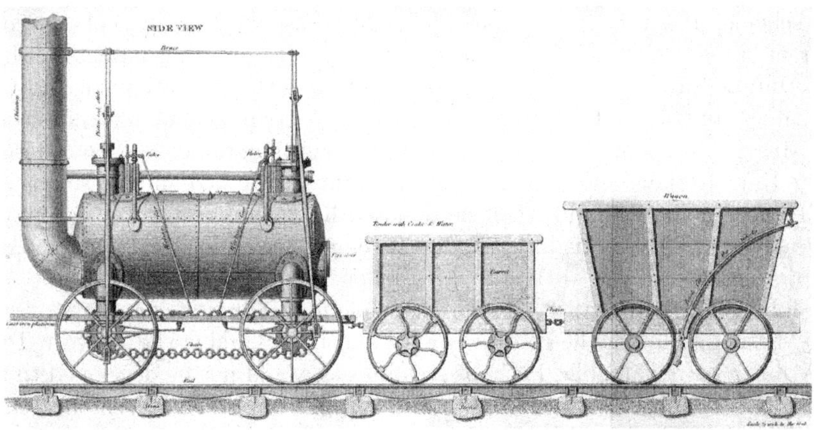

1830 American Locomotive

Nathan Read, the third of the steamboat patentees,[17] did not suffer from being an inventor. He went on, instead, to other things, becoming a congressman and a judge.

John Stevens Jr. of Hoboken, New Jersey, the fourth steamboat patentee, perhaps more of a steam engine inventor than a steamboat inventor in his earlier days, went on to practice engineering for the rest of his life, and left sons who were also engineers. He is said to have put the first steam-driven locomotive on tracks in this country in 1826. This was done on tracks on his

[15]See [How45] and R.C. Wakefield to George W. West, 12/26/1792 from NI 11/7/1816.

[16]See [How45].

[17]See [Rea70].

own estate just to prove it could be done. One of his sons founded the Stevens Institute of Technology, in Hoboken.

John Stevens' Private Railroad Track

It was not until years later, after the expiration of all of the 1791 patents, which occurred in 1805, that Robert Fulton obtained the financial backing to make the steamboat a practical proposition.

Musical Interlude

During the presidency of George Washington, when Oliver Ellsworth was Chief Justice of the United States, about 1796, Washington visited Ellsworth in his home at Windsor, Connecticut. While he was there, when the dinner hour arrived, the Ellsworths searched for the President. They found him in the upper chamber[18] with Oliver Ellsworth's five-year-old twin sons, Billy and Harry, one on each knee, bouncing them up and down while singing them a blood-curdling ditty called "The Darby Ram." Billy would become William W. Ellsworth, Governor of Connecticut. Harry would become Henry L. Ellsworth, the first Commissioner of Patents. Oliver Ellsworth would later resign the office of Chief Justice of the United States to take on the then more important job of Chief Justice of Connecticut.

[18] See [DotAR07].

The Derby Ram.

(Concluding Symphony as above.)

The Darby (Derby) Ram was originally an English folk song,[19] and was later adapted and modified to suit the American experience. There are so many different versions of the lyrics, and even minor variations in the tune, that it would be impossible at this date to be certain which version the President sang. About a hundred years ago, the song had been so thoroughly Americanized that it became a ragtime classic,[20] and later a jazz classic, called "Oh Didn't He Ramble."

One version of the tune and lyrics is given here:

THE DARBY RAM

As I went down to Darby, Sir, 'twas on a market day, I saw the finest ram, Sir, that ever was fed on hay.

Chorus: And indeed, Sir, 'tis true, Sir, I never was taught to lie, And if you'd been to Darby, Sir, you'd have seen him the same as I.

He had four feet to walk on, Sir, he had four feet to stand And every foot he had, Sir, did cover an acre of land.

Repeat Chorus

The horns that grew on his head, Sir, they grew up to the sky, And eagles built their nests there, for I heard the young ones cry.

Repeat Chorus

The wool that grew on his tail, Sir, filled more than fifty bags, You'd better keep away, Sir, when that tail shakes and wags.

Repeat Chorus

The tail that hung behind him, was fifty yards and an ell And that was sent to Darby, to ring the town church bell

Repeat Chorus

The man that killed the ram, Sir, was drowned, Sir, in the flood, The boy who held the bowl, Sir, was washed away by blood.

Repeat Chorus

The man that owned the ram, Sir, I think is very rich, And the boy who wrote this song, Sir, is a lying son of a bitch

Repeat Chorus

In the London of 1798, one of the most prominent theaters was the Theatre Royal Covent Garden. This was the first of a series of theaters on this site and was opened in 1732. The latest theater on the site is still in operation as the Royal Opera House Covent Garden. Joseph Shepherd Munden (1758–1832) was an actor at the Covent Garden Theatre from 1790 to 1811, becoming the most celebrated comedian of his day.[21] On April 20, 1798, he opened as Robin

[19]See [BMF93],

[20]See [Gil78].

[21]See [SSL94, v. 13, 1195–98] and [Hog68, pp. 2060–61].

in The Waterman. That evening, in connection with this play, probably in the intervals between scenes, he introduced a lively upbeat song entitled Patents all the Rage.

The introduction to the printed version of this song reads:

PATENTS ALL THE RAGE A New Comic Song Sung by Mr. Munden at the Theatre Royal Covent Garden with Universal applause Written & Adapted to the Well known Tune of PUSH ABOUT THE JORUM By James Payne Author of the Herefordshire Song &c &c Entered at Stationers Hall Price 1s London. Printed by Holland & Jones No 23 Bishopsgate Street

The lyrics and music for the song follow:

In every clime and at every time some fashions have had sway, And curious strange and simple things by turns have had their day, No wonder then in this great Town in such a polished age, Sir, When art and genius are combined that patents are the rage, Sir. Chorus: No wonder then in this great Town in such a polished age, Sir, When art and genius are combined that patents are the rage, Sir.

Our Clothes our physic and our Food, with many queer utensils, Must all be marked with Patent Stamps like warming Pans and Pencils, In mentioning the various Arts you'll think I'm ringing changes, We've Jacks and Grates and Kitchens too, and also Kitchen ranges.

Repeat Chorus

We have Patent fleecy hosiery will open every pore, And such ills as Gout and Rheumatism soon kick out of Door, We've also Pills to cure or kill, Perfumes to please your Noses, With Lozenges and currant drops, and Warrens Milk of Roses.

Repeat Chorus

In Paternoster Row, we have a Patent Book of knowledge, What pity 'tis not infused among our Blades at College, Then by Patent they could preach or pray, and wisdom ne'er lacking, Would shine like Boots and Shoes well blacked by Baily's Patent Blacking.

Repeat Chorus

We've a Patent Urn and a Patent Churn, with Candlesticks and Snuffers, But some are rude enough to call the inventors only puffers, Tho in execution Week now to prevent domestic wrangle, The Men may get the washing up with patent Mill and Mangle.

Repeat Chorus

Was Phaeton now with the Sun to run his course again Sir, With patent Harness Wheels and Whip, divinely made by Men Sir, His days work He with ease may do guide the Sun safe to Bed Sir, And light the World with patent Lamps, to shine forth in its stead Sir.

Repeat Chorus

Lord Chesterfield said to his Son, mind grace in all you do Sir, Even paring of your pretty nails, or buckling of your Shoe Sir, And when seated on a private seat, there leaving a deposit, That Business you may do with grace, in a patent water Closet.

Repeat Chorus

Then we have got true Patent Shot, with Gun powder and Gigs Sir, We've patent Cauls and patent Hair and Ladies patent Wigs Sir, There's patent Paste will lather in haste, Razors to please all Faces, But the most pleasing of these pleasing things, are pretty patent places.

Repeat Chorus

We've Medicines by Patent, in every Street now sold Sir, Which if you'd take you'd live as long as the Patriarchs of old Sir, But enemies to Physic at such blessings will be scoffing, And die just to enjoy the sweets of a good snug patent Coffin.

Repeat Chorus

Allegro
In
ev'-ry clime and at ev'-ry time some fash'ons have had sway and
curi-ous strange and sim-ple things by turns have had their day
No wonder then in this great Town, in
such a polished age Sir, when art and genius are combin'd that
patents are the rage Sir, No wonder then in this great Town in
such a polish'd age Sir when art and genius are combin'd that
pa tents are the rage Sir

Chapter 6

The Patent Law of 1793

The Patent Act of 1790 did not satisfy applicants, and it did not satisfy those who were appointed to administer it. Applicants did not like the strictness with which the Patent Board determined who should and who should not receive patents. There was no appeal from the discretion of the Patent Board. The members of the Patent Board had high positions in the Government and did not have sufficient time to spare to run the patent system. They were already reviewing patent applications at their lodgings after regular office hours.

Thomas Jefferson wrote a letter years later in 1813,[1] stating that the investigations of patent applications had required more time than the members of the Patent Board could spare from their higher duties, and that the applications had required a great deal of time to understand and treat with justice. In a 1792 letter to Dr. Hugh Williamson, the chairman of a congressional committee to revise the patent law,[2] he wrote that it distressed him to render crude and uninformed opinions on valuable and important rights without being able to devote the full attention that the opinions deserved. He wrote a bill to reform the patent law on December 1, 1791, and discussed it with Dr. Williamson.[3] Mr. Jefferson's bill would have eliminated all examination and made the Secretary of State the registrar of patents, with determinations of patent validity to be made by the courts. It does not appear that this bill in the form written by Mr. Jefferson was ever introduced in Congress. In early 1793, Rep. Alexander White moved to appoint a Director of Patents to run the patent system, but Rep. Samuel Livermore opposed the motion because it would require the young nation to appoint and pay another officer, instead of

[1] Thomas Jefferson to Isaac McPherson 8/13/1813, see [Lip03, v. 13]

[2] Thomas Jefferson to Hugh Williamson 4/1/1792, see [BG90, v. 23].

[3] A Bill to Promote the Progress of the Useful Arts, 12/1/1791, see [BG90, v. 22].

tacking patents onto the duties of an existing officer.[4] While Mr. Jefferson's bill was not introduced as written, some of its provisions were incorporated into a bill which was passed as the Act of February 21, 1793. It appears that the primary reason for the Act of 1793 was not the dissatisfaction of inventors but was the inability of the Patent Board to administer the Act of 1790 while carrying out their other duties.

The Act of 1793 went from the extreme of rigid examination to the opposite extreme of no examination at all. The Patent Board was abolished. The State Department was to register patents, and the courts were to determine whether the patents were valid. The State Department could not refuse to issue a patent if a proper description and drawing were submitted, a model was supplied, the proper allegations made in an affidavit, and the proper number of witnesses supplied.

Eli Whitney

Although the Act of 1790 made no distinction between the rights of citizens and aliens to obtain patents for their inventions, the Act of 1793 restricted the right to citizens. The Act of 1793 raised the fee for obtaining a patent from approximately five dollars to thirty dollars, which was an immense sum of money for a farmer or blacksmith to pay at that early date.

The first invention of great importance patented under the Act of 1793 was the product of Eli Whitney (1765–1825) of Massachusetts. Eli Whitney was the son of a frugal, small farmer, who intended that his sons should be farmers after him. Young Whitney was a skillful mechanic from his early teens, and between 16 and 18 he made good profits producing nails at the forge to replace those unavailable from abroad because of the revolution. He determined to get a college education and worked as a mechanic and school-teacher for years, until he was able to enter Yale College at the age of 24. He excelled at math and physics, and studied classics and literature from necessity.[5]

Upon graduation, he engaged himself as a tutor to a gentleman from Georgia. On the boat to Savannah, he met a fellow passenger, Catherine Greene,

[4] See [Gal34, cols. 853–860 at 854 (1/30/1793–2/4/1793)].

[5] See [Wyn50].

widow of Revolutionary War General Nathanael Greene. On arrival, he found that his intended job was no longer available. Mrs. Greene invited him to live in her home at Mulberry Grove, near Savannah, while he studied law, which was then his intention. He accepted the invitation, and made himself useful around the plantation through the use of his mechanical skills.

In the spring of 1793, Mrs. Greene had visitors from Augusta, Georgia, who were old comrades of her late husband. These gentlemen, Majors Brewer, Forsythe and Pendleton, were discussing the problems they were having with agriculture on their plantations. While sea island cotton, which grows well only in the coastal area of the state, has seeds which are easily removed, the upland cotton which grows in other areas of the state has seeds which are very difficult to remove from the lint. The visitors were discussing how much upland cotton they could profitably produce if there were an easy way to remove the seeds. Mrs. Greene suggested that her visitors discuss the problem with her resident mechanical genius. He agreed to try to solve their problem.

The first problem was to obtain some upland cotton with the seeds in it. He had never seen any. He visited Savannah and succeeded in obtaining some cotton, and returned to Mulberry Grove to experiment on it. He confided his intentions only to Mrs. Greene and to Phineas Miller, a New England gentleman who was then the tutor for Mrs. Greene's family, and who subsequently married Mrs. Greene and became Mr. Whitney's business partner. Whitney was given a workroom in the basement of the house in which he made a small model of his invention in about ten days. It took him much longer to make a larger, successful working model. The necessary materials were very difficult to find in Savannah. He had to draw out all the wire that he needed for his model, because, at that early day, no ready-made wire was sold in Savannah. Mr. Whitney's cotton engine, or cotton gin, as he conceived it comprised a wooden cylinder with iron wire teeth set about its periphery. During its operation, these teeth seized the raw cotton, and pulled it through the narrow openings in a row of adjacent iron straps. The openings were too narrow for the cotton seed to pass, and they were brushed into a receptacle below. A second cylinder, moving in the opposite direction, was provided with wire brushes to remove the now seedless cotton from the wire teeth.

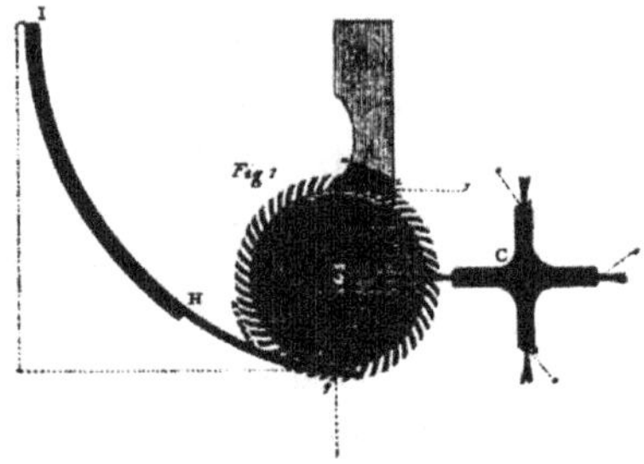

Original Drawing

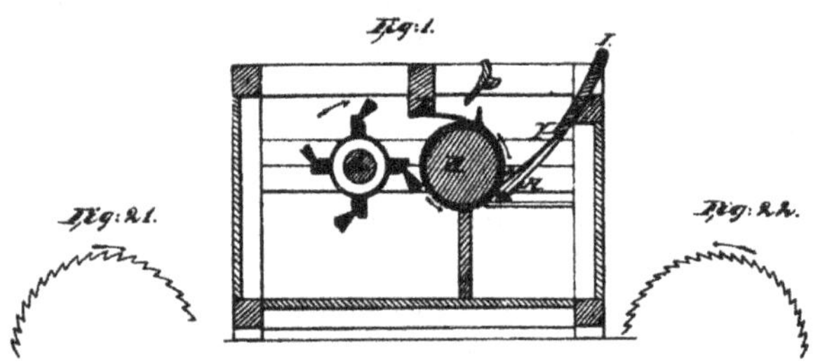

Restored Drawing

Patent Drawings for Whitney's Gin

Mr. Whitney returned to New Haven to confer with his friend Elizur Goodrich on the preparation and filing of a patent application. Mr. Goodrich was, at various stages of his life, a member of Congress, a collector of customs, a professor of law at Yale, and the mayor of New Haven. He was also the father-in-law of Henry Ellsworth, first Commissioner of Patents,[6] whom we meet later. Mr. Whitney applied for a patent on June 20, 1793, but did not receive it until March 14, 1794. While this may seem to be a relatively short time between application and patent when viewed by the present generation of applicants, it was an excessively long time under the old registration system.

Within a very short time, competitors were imitating and improving the cotton gin. The most successful improvement was patented in 1796 by Hodgen Holmes, of Augusta, Georgia. Mr. Holmes found the process of setting wire teeth into the wooden cylinder and maintaining them there to be unnecessarily complicated, and he replaced them with a stack of saw disks, the teeth of the

[6]See *National Intelligence*, 11/5/1849.

saws replacing the wire teeth of Mr. Whitney. This quickly became the only type of cotton gin in use. Mr. Whitney fought a long series of court battles to protect his invention, and it seems that he collected at least $90,000 in royalties over the term of his patent, which was a very large sum of money at that early date. In 1804, a certified copy of Mr. Whitney's patent was provided to the U.S. District Court in Savannah.

When the Patent Office burned to the ground in 1836, an effort was made to restore as many patents as possible. It is unclear where the Patent Office got the information to restore the Whitney patent, but the information was wrong. The restored patent shows Whitney's gin using saws, as all gins did by the time of the patent restoration. However, the certified copy, which was apparently not known to the restorers, shows that Whitney's gin used wire teeth.[7] The restorers can be pardoned for restoring Whitney's expired patent from insufficient information, since they did not know of the existence of sufficient information elsewhere. A good argument can be made that this one invention, by itself, has contributed more to the gross national product than the cost of running the patent system throughout its entire existence. Mr. Whitney went on in later life to help in the development of mass production, accepting and fulfilling a contract with the U.S. Government to produce ten thousand muskets.

It has been said that Whitney, by inventing the cotton gin, made slavery profitable in the South, and by developing mass production, made manufacturing efficient in the North, thereby doing more than any other person to insure Southern participation and Northern victory in the American Civil War.[8]

The Act of 1793 continued in effect, with only relatively minor changes, through 1836. The industrial revolution in this country was well begun under this act, with all of its shortcomings.

One of the notable inventors in the 1790s was Charles Willson Peale, the Philadelphia portrait painter, museum keeper, and father of a large family of painters named mostly for classical painters. On January 21, 1797, Charles Willson Peale received a patent for a bridge, the general structure of which is known from other sources. On November 16 of the same year, he received a patent for fireplaces. The next day, Eli Terry, founder of the American clock-making industry, received his first patent for a clock. On December 14, 1798, Raphael Peale, son of Charles Willson Peale, received a patent for preserving wooden wharves and vessels from worms. On June 28, 1800, Peter Lorrilard received a patent for a machine for cutting tobacco.

[7]See [Tom01].

[8]See [Co.57].

Sophonisba Peale, a painter and daughter of Charles Willson Peale, married Coleman Sellers, a Philadelphia manufacturer and son of that Nathan Sellers who was excused from the Revolutionary Army at the request of the Continental Congress to help make much-needed paper. During the presidency of George Washington, Nathan and Coleman Sellers had their shop in Philadelphia, across the street and down the block from the home of the President. A son of Coleman and Sophonisba Sellers, George Escol Sellers, will make his appearance later.

It may seem that there is a close connection between Philadelphia and invention. While this may be true, it is most probably because Philadelphia was the largest city in the nation at the time and the seat of the federal government. But in 1800, the federal government was about to move to an unsettled farming area, with a large proportion of swamp land, not far from the home of the late George Washington.

Chapter 7

The Government Moves to Washington

The infant federal government decided about 1791, in a political compromise, to move the federal capital south from Philadelphia. The Constitution allowed states to cede land to the federal government for use as the seat of government. A district was chosen, near the fall line of the Potomac River, which included the settlements at Georgetown, Maryland, and Alexandria, Virginia, as well as large areas of farm land and a generous allotment of swamps. The move was delayed for years, to allow time to erect some buildings in this wilderness for use by the federal government. Three Commissioners for the Federal City were appointed to superintend the initial building, allotment of land, and government of the federal district.

William Thornton

The Commissioners for the Federal City announced, on March 14, 1792, a competition for plans for the Capitol and the President's House. Plans by James Hoban for the President's House were accepted. Although numerous plans were submitted for the Capitol, all were objectionable. From Tortola in the British Virgin Islands came a letter from Dr. William Thornton (1761–1828), the same Dr. Thornton who had worked with John Fitch to develop the steamboat. Thornton was born on Tortola, but was raised by his Quaker grandmother and maiden aunts in Lancaster, England, and educated as a medical doctor in Scotland.

He had come to America in 1787 and became a citizen of Delaware in 1788. He married Anna Maria Brodeau of Philadelphia in 1790. He was on an extended stay in Tortola with his bride when he wrote to the Commissioners in October 1792, asking permission to submit drawings for the Capitol. His plan won the competition and formed the basis for the plans for the Capitol. The prize was $500 and a lot having a value of £100. He set up medical practice in Philadelphia and made his first visit to the federal district in March 1793, with a letter of introduction from President Washington. He declined to supervise the construction of the Capitol because of the time involved. Since Dr. Thornton had absolutely no formal training or hands-on experience as a supervising architect, he may have felt incapable; however, considering the number of fields that he dabbled in over his lifetime, he probably did not feel incapable of anything.[1]

Samuel Blodgett, Jr.

Washington, in 1793, was a "city" of about 300 residents, most of them speculating on the future of the proposed new capital city. One of them, Samuel Blodgett Jr., a native of New Hampshire and a Revolutionary War officer, had made a fortune in the East India trade and hoped to increase it in Washington. He promoted a lottery to advance his real-estate interests and offered a "Great Hotel" worth $50,000 as first prize. He built Blodgett's Hotel as the prize but did not finish it before he went bankrupt. Little is known of the early use of the partially completed structure except that some public meetings were held there. Meanwhile, the emerging city made do with the "Little Hotel" at Rhodes Tavern.[2]

A new set of Commissioners for the Federal City was appointed in 1794. Dr. Thornton was appointed a commissioner on September 2, 1794. His colleagues as city commissioners were Gustavus Scott and Alexander White. Dr. Thornton lived for a while in Georgetown, on what was then called Falls Street, but was more recently successively given the address 3221 Bridge Street and 3221 M Street. Since the name of the federal district was not yet settled, and Georgetown was no longer in Maryland, he sometimes gave his address as George Town, Columbia, and at other times as George Town, Potomac. In 1796 or 1797, he moved to Washington proper. He bought a city lot from

[1] See [Eva19]and [Cla15].
[2] See [Pro39].

Samuel Blodgett, probably complete with house, and he lived the rest of his life at what would become 1331 F Street, Washington.[3]

Dr. Thornton also purchased a farm, slightly larger than one square mile, on both sides of the road from Georgetown to Montgomery Courthouse (now Rockville), Maryland, located in what is now downtown Bethesda. He called it Park Grove, and it had a frame dwelling of one and a half stories, with two rooms on the lower floor and one on the upper floor. He also had a city garden located at the southeast corner of New York Avenue and Eighteenth Street, including twelve city lots, near the present Corcoran Gallery, for growing food for the table. He also had a 56-acre farm at Kalorama, just south of the present site of the National Zoological Park, where the Belmont Apartments now stand, for grazing his horses. He needed grazing land, because he and his next-door neighbor after 1800, Secretary of State James Madison, maintained a large stock of race horses and once started a racetrack to practice their hobby.[4]

After the preliminary work of setting up the seat of government had been done, the Government moved from Philadelphia to Washington in June 1800. The State Department had eight employees, and it was first crowded into a building erected for the Treasury Department. It is said, on unclear authority, that the State Department was moved temporarily by August 27 to one of the "Six Buildings," a group of buildings located from what is now 2107 to 2117 Pennsylvania Avenue. Numbered addresses did not come along for many years. By the end of 1800, the State Department was located at what is now 1901 Pennsylvania Avenue, where it remained until the building was burned in 1814. [5]

As the federal government moved to the new city, a theater company from Philadelphia arrived to present a play at Blodgett's Hotel on August 22, 1800. A room was fitted up for public performances, and the United States Theater presented Venice Preserved, or the Spoiled Child. The theater was permanently closed in less than a month.[6]

When Washington was chartered by Congress as a city with its own government on May 3, 1802, Thornton's appointment as Commissioner of the Federal City expired, and he needed a new job. [7] Although he once had a large private income, it was decreasing by this time. His job as Commissioner had paid $2,000 a year. Secretary Madison knew the State Department was then receiving more patent applications than could conveniently be handled by its other clerks. He thought Thornton could be in charge of the patent work.

[3]See [Cla15].

[4]See [Cla15].

[5]See [Pro42, B4] and [Cla15].

[6]See [Pro39].

[7]See [Cla15].

The total receipts for patent work were then about $1,400 a year, so he offered Dr. Thornton that sum as a salary, probably about May 12, 1802. The first record of Dr. Thornton's employment in the State Department indicates that he was paid for services rendered from June 1, 1802, to December 31, 1802.[8]

It is hard to look back and say exactly when a United States Patent Office, as opposed to a patent system, began. It did not begin with the passage of the Act of 1790, nor with operations under that act. Assigning three cabinet officers the duties of determining what patents should be issued did not establish an office dedicated to granting patents. Nor did it begin with the Act of 1793, under which various State Department clerks shared the work of engrossing patents between their diplomatic duties. But it is arguable that on June 1, 1802, with the appointment of a clerk whose entire and full-time job was superintending the issuance of patents, a United States Patent Office began. A small beginning, one clerk in one crowded room in a building devoted not only to the State Department but other departments as well, with no assistance and very little control over his own office—but a beginning.

And in late 1802, Daniel Brent was paid $22 for hay for the office pony. [9]

---

[8]See National Archives M234 roll 54 William Thornton account 1/6/1803; William Thornton to Henry Clay 12/21/1826, in H.R. Doc 47 19th Congress, 2nd session, Federal Documentary Microfilm, edition 1/4.

[9]See National Archives T903 roll 1 Brent account 10/9/1802

Chapter 8

Dr. Thornton Takes Charge

Because of the 1836 burning of the Patent Office, all that remains to tell us what happened in the early days is in scattered records that were held outside the Patent Office. As a result, some things can be seen well, some poorly, and some not at all.

Dr. William Thornton sent a letter from the Patent Office on June 9, 1803, to John Reid in which Dr. Thornton acknowledged receipt of papers and models for a patent application for an invention by John's brother James Reid for screwing and packing cotton.[1] Dr. Thornton agreed that John Reid would serve as his brother's agent for attending to the patent application. In 1807, Dr. Thornton had further correspondence with John Reid to suggest that his brother James should promptly execute a proper application for the patent, [2] the papers on hand being treated merely as a caveat. Caveats at this time were part of an informal procedure by which a person could file incomplete papers in the Patent Office to describe his invention while he perfected his invention. If a patent application for a similar invention was filed, the Patent Office would give notice to perfect the filing for purposes of an interference arbitration. An application for a similar invention had been filed by Richard and John Thorne, and Dr. Thornton thought the applica-

[1]See William Thornton to John Reid 6/9/1803 Federal Documentary Microfilm Edition No. 1/1

[2]See William Thornton to John Reid 8/4/1807 and 8/18/1807 Federal Documentary Microfilm Edition No. 1/1; John Reid to William Thornton 8/12/1807 Federal Documentary Microfilm Edition No. 1/1

tion should be perfected so that an interference could be declared and arbitrated. In further correspondence, it developed that James Reid had been in the Navy in the Mediterranean when the patent application was filed and had not returned to the United States in four years. The surviving correspondence with John Reid ends here, but James Reid never received a patent, while Richard and John Thorne received a patent for a cotton press in 1810.

William C.C. Claiborne

William C.C. Claiborne, formerly a congressman from Tennessee and the appointed Governor of the Territory of Mississippi, wrote a letter to Dr. Thornton on October 1, 1804, introducing Pierre Derbigny. [3] Mr. Derbigny was to deliver a little box containing a model of a new cotton machine for which the inventor, Obadiah Crawford, who was a citizen of the Mississippi Territory, solicited a patent. While Governor Claiborne stated that he was not personally acquainted with Mr. Crawford, it appears that he was acting as agent, and he requested that if the patent were granted, it should be forwarded to himself.

The first two records thus indicate that inventors prosecuted their patent applications through agents from an early date, and that the agent could be as close as the inventor's brother, or as highly placed as a territorial governor and former congressman. Indeed, it appears that many patent applications were brought to Washington and filed by congressmen for their constituents, although whether they usually received a fee or settled for votes is not apparent.

Dr. Thornton was also charged at an early date with registering copyrights. In 1804, he gave a certificate to William Pelham, [4] certifying that he had deposited with the State Department a copy of his publication Letters from London, in accordance with the copyright law.

The next reference we have to Dr. Thornton's work in the Patent Office relates to a patent which founded an American industry. In October 1804, Eleuthère Irénée du Pont de Nemours applied for a patent for a method of

[3] See William C.C. Claiborne to William Thornton 10/1/1804 Federal Documentary Microfilm Edition No. 1/1

[4] See William Thornton certificate 10/13/1804 Federal Documentary Microfilm Edition No. 1/1.

automating the granulating and sieving of gunpowder.[5] Mr. du Pont had established a factory near Wilmington, Delaware, to manufacture gunpowder by the best processes then known in Europe.

He recognized that the high cost of labor in the United States had always been a cause of discouragement of factories here, and recognized, as did generations of inventors after him, the uniquely American necessity of labor-saving inventions. In Europe, workmen were available in excess, and a working man had difficulty making enough by his work to support himself and his family. Thus wages were low, and thus men were cheaper than machines. Because working men were worried about keeping what work they had, they would fight the introduction of new labor-saving machines. But in the United States, there were few available working men and vast natural resources to harness. Any man capable of doing any work had a variety of options available and was generally not worried about labor-saving machines putting him out of work. The climate was favorable here for labor-saving inventions.

E. I. Du Pont

Mr. du Pont had invented a way to automate the granulating and sieving of gunpowder, and he was applying for a patent on his invention. One of his machines would do the work of six men, and he had erected four such machines, all of which could be kept in constant operation with the attendance of one man. The invention enabled the expansion of the du Pont gunpowder works, which later became the du Pont chemical empire. The patent application was written in French with an accompanying English translation. Peter Bauduy was handling Mr. du Pont's correspondence, perhaps as a secretary or associate of some type, perhaps another person acting as patent agent. In his reply to Mr. Bauduy, Dr. Thornton showed his wide range of interests by mentioning that he had written a dissertation on the manufacture of gunpowder, which he wished to send to Mr. Bauduy.[6]

[5]See du Pont to James Madison 10/24/1804 Federal Documentary Microfilm Edition No. 1/1

[6]See William Thornton to Peter Bauduy 10/27/1804 Federal Documentary Microfilm Edition No. 1/1.

Benjamin H. Latrobe

William Thornton was a friend from his Philadelphia days of Charles Willson Peale, the artist and the father of artists. In April 1805, Mr. Peale wrote Dr. Thornton to request a certified copy of a patent issued to John J. Hawkins in which Mr. Peale had an interest. The patent related to a physiognotrace and polygraph (apparently a device for tracing silhouettes), and Peale needed the certified copy to bring suit against a person who was making the device without authority. John J. Hawkins had been in England, where he sold patent rights to the polygraph and drawing machine for 1,600 guineas. Mr. Peale also wrote to Dr. Thornton in May 1805 to record an assignment of the Hawkins invention to Mr. Peale for the City of Philadelphia. It was common in the earlier days of the patent system to assign inventions by states, counties and cities.[7]

Benjamin Henry Latrobe, one of the successors to Dr. Thornton as Architect of the Capitol, and co-shareholder with him in a gold mine venture, wrote to Thornton in February 1806, applying for a patent for a new method of constructing stone bridges.[8] Latrobe stated that he had first used this method to build a bridge 15 years earlier. He applied for a patent at this late date only because he had submitted a plan for building a bridge by this method near Philadelphia, which was turned down, and others had later proposed to build two nearby bridges using his method. "Otherwise I should never have thought of obtaining a patent." There is no record in the list of issued patents that Mr. Latrobe ever received a patent for the invention, but it would not have been refused merely because it had been in use for 15 years. Dr. Thornton had no authority under the Act of 1793 to refuse patents even if copied without modification from patents previously issued to others.

On January 1, 1807, Secretary of State James Madison made a report to Congress, stating that the business relating to patents issued for useful arts had increased at the rate of doubling in four years. He recommended

[7]See Charles Willson Peale to William Thornton 4/16/1805 and 5/14/1805 Federal Documentary Microfilm Edition No. 1/1.

[8]See Benjamin Henry Latrobe to William Thornton 2/21/1806 Federal Documentary Microfilm Edition No. 1/1.

employment of an additional clerk.[9] He wrote that in 1806 William Thornton had been employed "in superintending and issuing patents for useful inventions and discoveries, in securing copyrights, etc." at an annual compensation of $1,400.

Senator William Plumer of New Hampshire kept a diary of events concerning his one term in office from 1803 to 1809. It is fairly clear that he did not like Dr. Thornton. He accused Thornton of superintending the finishing of the north wing of the Capitol in a such a shameful manner that he was deserving of censure, because the Capitol was constantly leaking and the walls in danger of falling. In fact, Dr. Thornton never supervised construction of the Capitol, although he did work closely with at least one of the many architects involved over the years during which he was a Commissioner of the Federal City. In the continuous battle between the successive Architects of the Capitol, Dr. Thornton's name did not escape without insult. On January 3, 1807, Senator Plumer visited Dr. Thornton in his office, not on patent business, but in connection with Dr. Thornton's secondary office as justice of the peace for Washington. It is clear that Senator Plumer was not impressed by Dr. Thornton or his office.

William Plumer

He wrote in his diary for that day: "This Justice Thornton is the keeper of the Patent office—who records all those inventions, and titles of books for whom patents and certificates of copyrights do issue. With him a set or volume of each book is lodged, and the model or drawing of each piece of mechanism, for which a patent has issued. His office is a room in the same building in which the War and Post office is kept. The floor and shelves are covered with models thrown together without any order or regularity. The books lie in an irregular confused pile on shelves and window stools covered with dust. The room is too small for the purpose; but a little money and labour would procure a convenient and useful book case and arrange the models and drawing in order. This Dr. Thornton ought to do—he has too long been guilty of great negligence." It should be noted that Senator Plumer did nothing to get Congress to provide Dr. Thornton with "a little money" to carry out his proposal, nor did he

[9]See James Madison to Congress 1/1/1807 Federal Documentary Microfilm Edition No. 1/1.

provide the clerk just recommended by the Secretary of State to provide the necessary labor to do more than just keep up with the doubling of patent applications in the four years of Dr. Thornton's stay in office. [10]

Dr. Thornton seems to have spent his spare time for most of his Patent Office career in writing letters to the Secretary of State and to Congress complaining of the poor way he was treated and the condition of the Patent Office. On November 27, 1807, Dr. Thornton wrote to Secretary of State James Madison to express dissatisfaction with his condition in the Patent Office.[11] He stated that he had troubled the Secretary often on the subject of patents without meeting with any encouragement. He claimed that the room which he then occupied was admitted by all to be inadequate. The Patent Office in Paris was said to have three rooms each as large as the whole Executive Office, and three directors with good salaries to superintend it, so important was it considered to be to the national prosperity. Dr. Thornton said that when he was engaged, he was expected only to superintend the granting of Patents and to carry on the necessary correspondence, etc., but by the time of this letter, when the patents were more than double in number, he was obliged not only to do these duties but also to write out the specifications in the issued patents, and to be the keeper of a museum. He requested an increase in salary to $2,000, a correspondence clerk, a proper office, an attendant, and the power of franking correspondence. A proper office was considered necessary for the full display of the models, and Thornton recommended that a house large enough for the purpose be rented at a cost of about $500 per year. The attendant would be necessary to preserve the models, keep them clean and in order, make fires, and serve as messenger. Dr.

Oliver Evans

[10]See [Plu23].

[11]See William Thornton to James Madison 11/27/1807 Federal Documentary Microfilm Edition No. 1/1.

Thornton considered that the institution would support itself and produce a surplus.

Remember Oliver Evans, the holder of the third U.S. patent granted under the Act of 1790. Mr. Evans' patent was for an improvement in the art of manufacturing flour and meal, granted December 18, 1790. The patent had been drawn up in the usual form of patents issued under that act, and expired in the usual 14 years, in 1804. Mr. Evans filed a petition for the extension of his patent rights, which was considered by a committee of the House of Representatives on December 7, 1807, after the patent had already been expired for nearly three years.[12]

Mr. Evans said that when his patent was infringed, he brought an action in the circuit court for the Pennsylvania district, where the court decided that the description given of his invention in the patent was insufficient, and his patent was therefore void. Since the sum in controversy did not amount to $2,000, he had no right of appeal to the Supreme Court. He petitioned Congress to grant him a good patent, and the committee understood that the only deficiency in the patent was a matter of form. Apparently, all patents were issued in the same form from 1790 until the passage of the 1793 act. Were all patents issued under the 1790 act invalid, based upon the decision of the Pennsylvania court with no appeal allowed? Mr. Evans had given a complete description of his invention in his petition to the Secretary of State in 1790, but, as a matter of form, the Secretary had not had the complete description incorporated into the patent as issued. Thus, it appears that the Pennsylvania court had found the patent invalid, not from any fault of Mr. Evans, but because the government was not incorporating the proper information available to it into the issued patent. When the committee asked the opinion of the Secretary of State, he was unwilling to recognize the invalidity of all issued patents based upon one unappealable court decision.

Oliver Evans' patent was extended by Congress, beginning a practice that continued for many years.

On April 11, 1808, New York Congressman Stephen Thorn wrote to Secretary of State James Madison[13] inquiring, more or less in the nature of a patent agent, about an application filed by his neighbor Jacob Coon for a machine for dressing timbers for wagon and carriage wheels. The practice of Congressmen representing inventors from their districts, and possibly from elsewhere, was also to continue for many years.

[12] See Report of Committee of House of Representatives 12/7/1807 American State Papers Miscellaneous 1:646 Federal Documentary Microfilm Edition No. 1/1.

[13] See Stephen Thorn to James Madison 4/11/1808 Federal Documentary Microfilm Edition No. 1/1.

On August 26, 1808, William Thornton wrote to Jacob Cist[14] of Philadelphia on a possible priority conflict between Cist and a Mr. Pike of Philadelphia. Mr. Pike had ceded priority to Mr. Cist. The invention related to a finely grindable material which could be used as black printing ink. On October 28, 1808, Dr. Thornton wrote Mr. Cist[15] regarding the delay in issuing a patent to Cist. Among other delays, Dr. Thornton had to await the signatures of the Attorney General, the President and the Secretary of State. Then he became ill and was confined to his bed for nearly a week, and to the house still longer. He also pointed out that in 1808 the post from Washington to Philadelphia only went out once a week. Dr. Thornton said that he thought the black pigment could make a fine Japan when used with varnish, and he proposed to add it to Mr. Cist's specification. It appears that Dr. Thornton paid the cost of getting the patent in exchange for an interest in it. Ethics for Government Employees was a very short course in those days.

On December 31, 1808, Thornton wrote again to Jacob Cist,[16] concerning the pigment patent. Mr. Cist had complained that the patent would have been better if the specification had been written in accordance with the specification submitted by the inventor, instead of as revised by Thornton. Thornton stated that he would alter any patent if he had made changes to the specification that the inventor did not agree with. He also stated that the same kind of coal, apparently meaning the coal which would be ground up for the pigment, had been found in Rhode Island, and suggested that it be included in the specification. The prohibition against introducing new matter into a patent application had not yet appeared.

Jacob Cist

On June 3, 1809, Thornton wrote again from Washington to Jacob Cist,[17] the ink inventor, stating that he had written to Mr. Perigrine Williamson, the inventor of steel writing pens with three slits, suggesting use of Mr. Cist's ink to prevent corrosion of the pens.

[14]See William Thornton to Jacob Cist 8/26/1808 Federal Documentary Microfilm Edition No. 1/1.

[15]See William Thornton to Jacob Cist 10/28/1808 Federal Documentary Microfilm Edition No. 1/1.

[16]See William Thornton to Jacob Cist 12/31/1808 Federal Documentary Microfilm Edition No. 1/1.

[17]See William Thornton to Jacob Cist 6/3/1809 Federal Documentary Microfilm Edition No. 1/1.

On September 1, 1809, Thornton wrote from Washington to Jacob Cist about the black ink.[18] Mr. Cist had asked Dr. Thornton what he valued his share at, presumably a share of the patent on the black ink. Thornton asked Cist for his aid in valuing his share. Thornton also mentioned that he had applied to the Navy Office, apparently in an attempt to sell the ink or to have it approved for Navy use. Thornton thought it more useful as printer's ink than as anything else.

William Thornton enjoyed his farm in what is now Bethesda. He kept some of his race horses there, and spent as much time there as possible. On December 17, 1808, he wrote to Secretary of State James Madison from the farm.[19] He said that he did not wish his reasons for visiting his farm at present to be generally known but felt that he must inform the Secretary of the reason. He had been very unfortunate in the recovery of debts due to him and had lost many thousands of dollars in the hands of persons whom he trusted. He had also become bail bondsman for Samuel Blodgett, the builder of Blodgett's Hotel, who was being held for debt, and when Blodgett fled, Thornton remained responsible for more than $10,000. He said he had recently sold the only house he had in Lancaster, England, to pay debts. He had attempted to mortgage his farm. He had received no money from his property in the West Indies for two years. He retired to the country until the court met, which required him to stay at his farm for a week longer. He had brought patent applications to his farm to write up on parchment for patents. He said he would issue 160 patents by the end of the year, which would amount to four times as many per year as when he first accepted the duties. He also said that he was unwell both in body and in mind when he came to the farm, and was not well at present.

State Department, Washington, D.C., Home of the Patent Office, 1801–1810

On May 7, 1810, Thornton wrote from his farm to Thomas Jefferson, commenting that he was at his farm for two or three days, for the first time in six

[18]See William Thornton to Jacob Cist 9/1/1809 Federal Documentary Microfilm Edition No. 1/1.

[19]See William Thornton to James Madison 12/17/1808 Federal Documentary Microfilm Edition No. 1/1.

months.[20] He indicated that his work had increased sixfold since taking the job, having issued 219 patents the previous year. He said that he was without any assistant, except when he hired one at his own expense. He thought some of the inventions did honor to the country.

[20]See William Thornton to James Madison 12/17/1808 Federal Documentary Microfilm Edition No. 1/1.

Chapter 9

The Feud Between Dr. Thornton and Robert Fulton

William Thornton worked with John Fitch on his steamboat back in the late 1780s and early 1790s, and he was very protective of the rights of the late John Fitch to his steamboat and of his own rights to improvements which he invented then and later. When Robert Fulton began work on steamboats and approached state legislatures to secure from them the exclusive right to work steamboats within their boundaries, William Thornton was incensed. He worked with whatever allies he could find to thwart Mr. Fulton at his every attempt to monopolize what Thornton considered common property in steamboats as invented earlier by Fitch and others.

On November 24, 1808, Dr. Thornton wrote to Colonel John Stevens, who was also building and operating steamboats.[1] Thornton informed him that he had written to New Orleans to warn against granting Robert Fulton an exclusive right to navigate the Mississippi River with steamboats. He also said that he had met with Mr. Fulton and mentioned the subject of the patent for the North River to him. After Fulton rebuilt a boat named the Clermont, he had renamed it the North River. Thornton was of the opinion that Fulton was preparing to take out a patent for a steamboat. He noted that Fulton had not yet taken out any patent whatever, and that, therefore, Thornton could

[1]See William Thornton to John Stevens 11/24/1808 Federal Documentary Microfilm Edition No. 1/1.

say nothing more than that the earliest information would be given to Stevens when Fulton applied.

On January 11, 1809, Colonel Stevens wrote to William Thornton[2] from New York, asking for a copy of Fulton's patent when it was taken out. Thornton replied[3] on January 23 that Fulton had just applied on January 19. Thornton had taken out a patent of his own on January 16 for some of his own improvements on steam engines. Thornton told Stevens that he would determine if Fulton would permit a copy of his patent to be given out, and, if so, he would inform Stevens and have a copy made.

Robert Fulton

Stevens wrote back to Thornton on January 28, requesting a copy of Thornton's patent, and also expressing surprise that his ability to get a copy of Fulton's patent could depend upon permission from Fulton.[4] Thornton wrote to Stevens on February 15, stating that Fulton had declined to allow a copy of his patent to be given out.[5] Thornton said that it had not been the practice of the Patent Office to give out copies of patents without permission of the patentee except in cases where suits in law had rendered it necessary. Thornton felt that he could not, with propriety, send Stevens a copy of Fulton's patent, but stated that if Stevens desired it, he would lay the request before the Secretary of State or Attorney General and act accordingly. Thornton said that the Secretary of State had decided such a question in favor of an applicant that same

[2]See John Stevens to William Thornton 1/11/1809 Federal Documentary Microfilm Edition No. 1/1

[3]See William Thornton to John Stevens 1/23/1809 Federal Documentary Microfilm Edition No. 1/1.

[4]See John Stevens to William Thornton 1/28/1809 Federal Documentary Microfilm Edition No. 1/1.

[5]See William Thornton to John Stevens 1/15/1809 Federal Documentary Microfilm Edition No. 1/1.

day, and that it might become a general rule. He also said that with respect to his own patent, he had no objection whatever to furnishing a copy.

On February 17, Horace Binney wrote to John Stevens from Washington,[6] stating that he had visited Dr. Thornton to request a copy or an abstract of Mr. Fulton's specification. Thornton informed Mr. Binney that, by a recent direction of the Secretary of State, copies of patents were to be given upon request in cases of this kind, and that accordingly he would have a copy made for Stevens. He said that it would be a work of some time, because of the press of business in the office, as well as the length of the specification. Binney wrote that Thornton had told Fulton that his patent was not worth sixpence. On February 21, Nicholas King gave Colonel Stevens' agent a receipt for $12 for making a copy of Thornton's steam engine patent.[7]

On May 12, William Thornton wrote to Robert Fulton,[8] indicating that Colonel Stevens had asked for a copy of Fulton's patent, but that Thornton was unwilling to supply it contrary to Fulton's wishes until the Attorney General answered the Secretary of State's request for an opinion as to whether it should be supplied.

On July 8, Dr. Thornton wrote to Colonel Stevens,[9] stating that he had been instructed by the Attorney General that he should furnish copies of patents to those who requested them. He wrote that if Colonel Stevens requested a copy of Fulton's patent, it would be furnished. On July 28, Colonel Stevens wrote to Thornton,[10] declining a copy, stating that Fulton was no longer threatening him with patent infringement, even though he was running his boat on the Delaware between Philadelphia and Trenton. Stevens also indicated that Thornton had offered to make him a partner under Thornton's patent, which offer Stevens declined.

In his May 12 letter to Fulton, Thornton went into some detail about why he considered the subject matter of Fulton's patent to be old. He said that the use of wheels on the sides of a boat was proposed at the time of Isaac Newton and published in the Lexicon Technicum nearly a hundred years ago. Thornton also said that he patented the wheel at the stern of the boat before he knew Fulton had any idea of such an invention. Thornton stated that Robert Fulton's description of proportions of his boat could not be the subject

[6]See Horace Binney to John Stevens 2/17/1809 Federal Documentary Microfilm Edition No. 1/1.

[7]See Nicholas King receipt 2/21/1809 Federal Documentary Microfilm Edition No. 1/1.

[8]See William Thornton to Robert Fulton 5/12/1809 Federal Documentary Microfilm Edition No. 1/1.

[9]See John Stevens to William Thornton 7/28/1809 Federal Documentary Microfilm Edition No. 1/1.

[10]See John Stevens to William Thornton 7/28/1809 Federal Documentary Microfilm Edition No. 1/1, wherein it is quoted.

of patent rights, because the law expressly excluded different proportions or mere variations in form from being patentable improvements.

Thornton seems to have been attempting to force a business connection between himself and Fulton. He suggested that Fulton read Thornton's patent, and he suggested the possibility of a connection with Fulton, in which he would give his aid in the prosecution of the business. Most peculiarly, Thornton stated that the Lexicon Technicum he referred to was in his possession and that he had kept its contents secret from those with whom Fulton was contending.

Thornton stated that the proportions proposed by Fulton were exactly the same as were used in the boat used by John Fitch's company in which he held shares many years ago. "Our boat was exactly eight feet wide and sixty feet keel—yours twenty by 150, each equal in length to 7 1/2 times the breadth! We also used a machine on Watt's and Boulton's principles. The cylinder I had prepared for a boat, intended for the Mississippi, of 25 tons burthen, was 3 feet 6 inches diameter, 4 foot stroke, the piston to work 30 single strokes a minute. All the proportion of force and velocity depend after this on the quantity of steam supplied by the boiler, for the size of the paddles or buckets of the wheel may be varied with little expense accordingly, but will forever vary according to the force or high power of the steam, which may be varied ad infinitum"

On January 29, 1810, William Thornton wrote to the Speaker of the Senate of Ohio, among others. [11] He opposed an attempt by the Fulton interests to obtain from the legislature of the state exclusive steamboat navigation rights on the waters in the state. He said that this would effectively deprive many ingenious inventors and mechanics of the benefits expected from the patents which they had taken out. It would also deprive the public of the free use of inventions which they should have after the expiration of the patents.

William Thornton also wrote a similar letter[12] to the Virginia legislature in 1810, opposing the grant of an exclusive steamboat navigation right on the waters of Virginia. The letter was laid before the Speaker of the House of Delegates after he had put the question of the third reading of the bill. The vote had not been taken, but the members were prepared to pass the law without any objection. The moment after the Speaker read Thornton's letter, he put the question again, and the vote was unanimously against the bill.

About January 1811, William Thornton wrote a similar letter to the legislature of the State of New York.[13] He suggested that the rapid progress of the

[11] See William Thornton to Speaker 1/29/1810, printed in Ohio Senate Journal Federal Documentary Microfilm Edition No. 1/1.

[12] See [Tho14, enclosure 2].

[13] See William Thornton to New York Legislature ca Jan 1811 Federal Documentary Microfilm Edition No. 1/1.

useful arts in England dated from the time when patents were first granted. He stated that the American people made a constitutional provision for granting patents in order to achieve the same results, and that the granting of patents was a federal function, which should be understood to preempt the state's right, formerly exercised, of granting patents of its own. The ability of each state to grant patents would make the federal goals impossible to achieve. Congress required patent owners to relinquish previous state patents in order to obtain corresponding federal patents. This clearly suggests that states should not now be granting new state patents. Thornton requested New York to reconsider its earlier grant of a monopoly for Livingston and Fulton to navigate the rivers of New York by steam, because John Fitch had a federal patent on a steamboat in 1791. When that patent expired, the public obtained the right to navigate by steam anywhere in the United States.

Fulton replied that while the United States may grant to an individual a patent to his invention, that simply means that no one shall use the patented invention without his consent.[14] The United States cannot authorize him to use his patent where he pleases. One may have a patented bridge and raise a company to build it, but one cannot build it over any river in the Union without the consent of the state through which the river runs. The state of New York can grant an exclusive right to a ferry over the North River for any number of years, and the owner of a patent on a steamboat can not use that boat on the ferry without the permission of the owner of the right to operate the ferry. The state has, he said, granted to Mr. Livingston and Mr. Fulton the right to operate a ferry from New York to Albany and, regardless of the patents which may be granted for steamboats, the patented steamboats cannot be navigated over their route without their consent.

On November 29, 1811, Robert Fulton wrote a nominally polite but obviously threatening letter to Dr. Thornton,[15] apparently demanding that he sign a previously provided "certificate which is an honest statement of facts" in order to "correct in the public mind an infinite injury" said to have been caused by Dr. Thornton's statements. Fulton threatened "more rigid measures" if this were not done. Fulton stated that it was not appropriate for the director of the Patent Office to volunteer writings and opinions on Fulton's United States patent or state grant. This, he said, was only for the United States courts and those whose evidence was requested. He also objected that Dr. Thornton had been attempting to force himself into a partnership with Fulton. Fulton

[14]See Robert Fulton reply 1/20/1811 to letter of note 13, Federal Documentary Microfilm Edition No. 1/1.

[15]See Robert Fulton to William Thornton 11/29/1811 Federal Documentary Microfilm Edition No. 1/1.

also said that after he had proved the superiority of water wheels for taking a purchase on the water, Thornton patented Fulton's invention of a wheel behind a boat, and then asked Fulton for $5,000 per year to cede to him his own inventions. Although Thornton cited Fitch's steamboat to Fulton, Fulton maintained that his specific invention had no resemblance to Fitch's boat. He said that comparisons with dormant and unproductive ideas could not be used to invalidate a patent, for otherwise a host of abortive ideas would destroy any patent and no useful inventor could be safe. Fulton requested that Thornton call on him to render all the justice in his power in order to prevent further trouble.

James Monroe

Apparently, Thornton did nothing or not enough to satisfy Fulton, because on February 13, 1812, Robert Fulton wrote to Secretary of State James Monroe to complain of Thornton.[16] He stated that Dr. Thornton was involving inventors in endless embarrassments and should be removed from the Patent Office. There were all sorts of restrictions on the dealings of various government officials to protect them from conflicts of interest, but, said Fulton, Dr. Thornton took patents of his own and then disputed the rights of those who interfered with his pretensions.

Fulton remarked that in January 1807 when he saw Thornton in Washington, Thornton told him that 18 years earlier in company with Mr. Fitch he had made a steamboat run eight miles an hour in still water. However, Fulton observed, such a speed is impossible and is sufficient proof that Thornton did not understand the principles which govern the construction of successful steamboats. Thornton, however, had not taken out a patent for his pretended portion of the invention in eighteen years or made any effort to put it in practice, so that the nation was not benefitted by the invention even if he made one. In the patent of Mr. Fitch, Dr. Thornton's name is not mentioned. Fulton said that in 1807 he set his first boat in motion on the North River with a wheel on each side of her, and the boat ran that autumn and the season of 1808. Still, Dr. Thornton did not take a patent for his pretended inventions. In the autumn of 1808, Robert Fulton came to Wash-

[16]See Robert Fulton to James Monroe 2/13/1812 Federal Documentary Microfilm Edition No. 1/1.

ington, and while there he began the drawings, specifications and descriptions for his patent, which took him nearly three months to complete. He said that he informed Dr. Thornton he was preparing the patent application, and at that time, one month before Fulton was ready to file his application, Thornton took out a patent for himself. This was eighteen months after Fulton's boat had been running on the North River, and obviously, in Fulton's opinion, was done to block him in using his own invention.

Fulton told Monroe that Thornton then offered to enter partnership with Fulton and Livingston for equal thirds. When this proposal was rejected, he offered his assistance for one eighth of the net profits. When this second offer was rejected, Thornton wrote two letters to prove that Fulton's patent and state grant were not legal. Fulton maintained that this action by the head of the Patent Office encouraged others working in this field to infringe his rights and involve him in a ruinous lawsuit. Thornton made another attempt to force himself into partnership in January 1811. Fulton provided Monroe with a copy of a letter in which it appears that Thornton proposed to protest to the New York legislature against monopoly grants to Fulton unless he were taken into partnership. When Thornton's offer was not accepted, he told an acquaintance of Fulton that he intended to do everything in his power to destroy Fulton's patent.

Fulton asked Monroe to make Thornton sensible of his duty. saying: "His business is perfectly neutrally not to interfere with patentees or give opinions. Leave that to the law. If he has genius for inventions and can live by it, he has a right so to do." But he has no right to take out patents for himself, deal in patents, and be director of the Patent Office.

Almost three years later, on December 27, 1814, Robert Fulton wrote another letter on the same subject to Secretary of State (and of War) James Monroe.[17] The only major difference in the substance of the letter was to the effect that Dr. Thornton had anticipated another of Fulton's invention, to the great injury of Fulton and the grievous injustice to the useful arts and artists of the country.

Immediately, on December 27, 1814, Secretary of State Monroe proposed a regulation to prevent the Superintendent from having an interest in patents, beginning the following February, in accordance with similar restrictions placed on other officers of the government.[18] Dr. Thornton replied[19] that it seemed

[17]See Robert Fulton to James Monroe 12/27/1814 Federal Documentary Microfilm Edition No. 1/1.

[18]See James Monroe to William Thornton 12/27/1814 Federal Documentary Microfilm Edition No. 1/1.

[19]See William Thornton to James Monroe 1/9/1815 Federal Documentary Microfilm Edition No. 1/1.

that Fulton's threats had succeeded in preventing him from having a right that every other citizen of the United States had. He pointed out that he was given a salary of only $1,400 per year, and thus had to have outside income in order to live, while the other officers thus restricted were given salaries of up to $5,000 per year and had no such necessity. Other officers were prohibited certain interests because they were entrusted by the government with secrets which would give them an unfair advantage over their fellow citizens. But the Superintendent of the Patent Office could have no such unfair advantage, because a patent application, when filed in the Patent Office, was dated and entered. If someone were to try to take advantage of an opportunity to read such a document, it would be apparent immediately. The Superintendent had no authority to refuse a patent except for lack of citizenship, insufficient witnesses, or nonpayment of fees, and could not affect the right of an applicant to obtain his own patent. Soon afterwards, on January 9, 1815, Fulton wrote to Attorney General Richard Rush, asking what had been done about his complaint to Secretary of State Monroe.[20] He wrote that Thornton was causing him great harm, and that, while asking nothing for himself, he wanted Thornton to burn all his patents if he was to continue in office. Otherwise, he intended to go to law against Dr. Thornton.

James Monroe wrote a note[21] in January 1815, mentioning his refusal to sign a patent to Dr. Thornton the previous December, apparently beginning to apply the prohibition well before February 1815. However, it appears that this prohibition was withdrawn, since the records indicate that Superintendent Thornton was granted a patent for paddle wheels on steamboats on December 27, 1814, and he received one additional patent on July 31, 1827, less than a year prior to his death.

In February 1815, Dr. Thornton wrote to Benjamin Henry Latrobe, who was also interested in steamboats. Latrobe wrote back,[22] thanking Thornton for information provided about the possible invalidity of Fulton's patent. But, Latrobe observed, Fulton had never been willing to rely on federal patents, and would only build steamboats where he had previously obtained a state grant independent of patents. Latrobe said that no one was as well acquainted with the details of Fulton's maneuvers as he was. He said that he was somewhat ashamed that at his age and with his experience he should so blindly have

[20]See Robert Fulton to Richard Rush 1/9/1815 Federal Documentary Microfilm Edition No. 1/1.

[21]See James Monroe to John Graham 1/26/1815 Federal Documentary Microfilm Edition No. 1/1.

[22]See Benjamin Henry Latrobe to William Thornton 2/13/1815 Federal Documentary Microfilm Edition No. 1/1.

confided in Fulton, and that he "had not sufficient penetration to see under his exterior" to "the selfishness" and "low cunning of his character."

Blodgett's Hotel (upper right) from 1803 watercolor

Robert Fulton died suddenly on February 24, 1815. In July 1817, Superintendent Thornton wrote to Attorney General and acting Secretary of State Richard Rush, giving his version of the controversy.[23] Thornton said he had evidence that Fulton had taken his ideas for steamboats from papers which John Fitch took to France in the early 1790s. Thornton printed a pamphlet in 1812 describing the work of Fitch and its relation to the work of Fulton.[24] Thornton sent a copy of the pamphlet to Fulton, and said that Fulton swore vengeance and vilified him to members of Congress until Thornton sent him a message via a friend that he would cut off his ears if he mentioned his name again. Thornton told Fulton to set forth his complaints, sign them, and send them to the Secretary of State, who could be counted on to do justice. The Secretary endorsed him to Congress, and Fulton declared that one of them must quit the public service. Thornton promised that in three months Fulton would not be able to show his face in public. Fulton had all of his friends join in the attack against Thornton. Thornton declared that he wrote something

[23]See William Thornton to Richard Rush 7/12/1817 Federal Documentary Microfilm Edition No. 1/2.

[24]See [Tho14] and William Thornton to William Duer 6/22/1818 Federal Documentary Microfilm Edition No. 1/2.

to Fulton and sent it to him in New York, that he read it with symptoms of horror, and died a few days later. Thornton said that Fulton had sought first to have him fired, but as he "had no salary worth considerations," he decided to prevent him from taking out patents and thus have him resign. Thornton said that the Secretary allowed him a period of time to take out patents on his prior inventions, after which he would be forbidden to take out patents. He had in 1817 never taken another patent because he had no money to pay for them. Thornton had appealed to the President to have this limitation on his right to take out patents revoked, and the President had referred the matter to the Attorney General for decision. In the July 1817 letter to Richard Rush, Thornton asked for a favorable decision from Rush as the Attorney General and acting Secretary of State.

Chapter 10

The Patent Office Finds a Home of its Own

By the Act of April 28, 1810,[1] Congress authorized the President to purchase a building suitable for use by the Patent Office and the General Post Office. President James Madison exercised this authority by buying Blodgett's Hotel, begun in 1793 by Samuel Blodgett and then still unfinished. The building and surrounding lots were purchased for ten thousand dollars from Robert S. Bickley,[2] who had won them in Blodgett's lottery.[3] Blodgett's Hotel was located on the north side of E street N.W., between what would become 7th and 8th streets. A block north was the Ridge, or F street, which was the main highway from Georgetown to Upper Marlboro.

On June 12, 1810, Benjamin Henry Latrobe, the architect, wrote to Secretary of State Robert Smith, indicating that he had examined the building and was offering his report.[4] All of the internal and external walls of the building were finished and were considered of excellent material and workmanship. The only deficiency was that the stacks of the south chimney were not carried through the roof. The floors consisted only of joists, which were of inferior quality. Some joists had apparently been removed for use as firewood by the squatters who had been living in the cellar of the building. The roof was tight and needed only to be filled in where the south chimney would be taken

[1] See 11th Congress 2nd session, Stat 2 Chap 34.

[2] See Albert Gallatin to Benjamin Tallmadge 2/6/1812 Federal Documentary Microfilm Edition No. 1/1.

[3] See [Pro39].

[4] See Benjamin Henry Latrobe to Robert Smith 6/12/1810 Federal Documentary Microfilm Edition No. 1/1.

through. Sashes and window frames were on hand for most of the windows, although all the glass had been destroyed.

Blodgett's Hotel, 1810–1829

Latrobe recommended that the building first be cleared of the three families that inhabited the cellars and one room at the west end of the ground floor. They kept hogs, and committed nuisances in the house which rendered it offensive in the highest degree, apparently Latrobe's euphemism for the use of the inside of the building as a privy for the inhabitants. He understood that the families were not so indigent as to require assistance, and, in fact, that they were all related and one of them owned valuable tracts of land in the city. After the families were removed, Latrobe recommended fencing the lots in the rear and on the east end to prevent further squatters. Then the land should be ditched and drained to prevent rainwater gathering in the cellars, and the cellars should be cleaned of the filth collected in them.

Latrobe recommended that the Patent Office be given the northwest corner room on the ground floor (a small room close to the west room) for the messenger. The southwest room on the first floor (which Americans would now call the second floor) would be used as an office for the Superintendent of the Patent Office, and the large room extending from the north to the south front through the center of the building would be used for the exhibition room and deposit of models. To render the latter rooms acceptable, it would be necessary to floor the ground story as far as the staircase, to carry the flooring up the stairs, and to floor the passage above.

He suggested this plan in preference to one proposed by Thornton, because it preserved the whole of the east end of the building, which had separate stairs, for the use of another public office which might also use the building, and it reserved the south room at the east end for such office. Otherwise, if the whole south front were occupied by the Patent Office, only the inferior north rooms would remain for a future establishment. Another objection to Thornton's plan was that it required a long common passage from east to west through the whole building. Thus, the center of the building would be lost to every other use, to the great annoyance of all occupants

Latrobe proposed to floor and counter-floor the northwest corner room, the passage above stairs, the whole of the center of the house upstairs, and the southwest room. He would support the floor of the exhibition room by eight columns, and would support the ceiling of the exhibition room in the same way. He would then put in the windows and wall up behind three superfluous windows on the west front. He would put in a plain wide staircase from the ground floor to the first floor and frame and put in all doors. He would carry the chimney stacks on the south front through the roof and make the roof good around them. He would put in baseboards and plaster the rooms. Lastly, he would put a framed privy in the yard

Robert Smith

He estimated that it would cost $1,820 to do the work, but recommended some latitude be included, and $2,250 be considered the upper limit of the estimate. This did not include any shutters, cases for books or models, grates or stoves, or any other form of furniture.

On September 3, 1810, Benjamin Henry Latrobe wrote to William Thornton, indicating that the new Patent Office was then so far finished that models might be moved there without inconvenience.[5] Latrobe suggested the following September 6 as the day when Thornton could take possession of the new office. He said that the messenger's room below and the southwest room upstairs were furnished as well as the available funds would admit. The southwest room would hold all of the models then in the Patent Office, although not arranged in order, and they could be taken from there into the model exhibition

[5]See Benjamin Henry Latrobe to William Thornton 9/3/1810 Federal Documentary Microfilm Edition No. 1/1.

room when that room was finished. The Secretary of State apparently wanted the space then used by the Patent Office evacuated that week. Latrobe wrote that the model exhibition room would be completed when Thornton told a Mr. Lenox (apparently a cabinetmaker) his ideas as to the kinds of cases or shelves he wanted, and when the Secretary of State arranged to fund the work. Latrobe suggested that a person be appointed to sleep in the building, and that if appointment of a messenger were to be necessary, Thornton should immediately approach the Secretary.

In the same letter, Mr. Latrobe indicated that he was preparing a patent application for Mr. Samuel Blydenburgh for a loom, which would soon be transmitted to the Patent Office. Benjamin Henry Latrobe, noted architect, was also a part-time patent agent.

On September 5, Latrobe wrote to Secretary of State Robert Smith,[6] stating that the sum of $240 would be necessary to make and put up the stands and cases for models in the new Patent Office. Without this furniture, he predicted that much confusion and destruction of the models would take place. In December, Latrobe wrote Thornton concerning the finishing work for two model cases that had been installed and suggesting that Thornton should have a closed case in his office as the most convenient means of keeping private papers out of sight.

Dr. Thornton set out to make his life easier by getting the government to employ a messenger for the Patent Office. He wrote to Secretary Smith about the condition of the Patent Office.[7] He began by discussing the extensive nature of the patent departments of France and England, and harking back to the time when the U.S. patent system was considered sufficiently important to claim the joint attention of the Secretary of State, the Secretary of War, and the Attorney General of the United States to fulfill a duty that was thought to require more discrimination and judgment than time or labor.

He indicated that when a patent was demanded, he was in the habit of doing a cautious examination of all that had been done in the same line before the receipt of the application. He said that the examination could not be satisfied by indices, nor by models, nor even always by referring to the papers and drawings, because it required an understanding of underlying principles. These principles are not always comprehended by ingenious but unlettered mechanics. He recognized that these examinations were not required by law and were more duties of conscience than of office. He knew that his resulting cautions

[6]See Benjamin Henry Latrobe to Robert Smith 9/5/1810 Federal Documentary Microfilm Edition No. 1/1.

[7]See William Thornton to Robert Smith 12/19/1810 Federal Documentary Microfilm Edition No. 1/1.

were often not received with thanks by the applicants, but he considered it proper to guard against deceptions, in which inventors may begin by deceiving themselves as to the originality of their inventions, and end by deceiving their fellow citizens. He could not refuse a patent for lack of novelty, but could point out that lack.

He wrote that the duties of the office also included particular attention to the necessary forms of patents. In consequence of a defect in the forms, many of the patents issued during the last century were set aside in the courts of law. Where applications appear to involve the same principles, reference must be made to arbitrators to decide on the priority of the invention. He also paid attention to the state of the papers and drawings to endeavor to render them worthy of the archives of a great people. All these duties require a large and difficult correspondence. Books were to be kept which register all these, and the books that are published in the United States are sent to be entered in the Patent Office, each of which requires a certificate. He said that the work had become so extensive that he had been obliged to hire an assistant from his private purse before the Secretary of State furnished him with an assistant. He also said that notwithstanding the large number of patents which he was obliged to prepare for issue, the patents which he issued gave him less trouble than those he refused.

He said that many of the members of Congress had witnessed the astonishing inventions of our countrymen, and that Congress, finding so many valuable models sent to illustrate the inventions which now enrich the Depository of the Arts, had purchased a spacious building for their accommodation. Thornton then got to what was apparently the purpose of his letter and suggested engaging a doorkeeper or messenger, who would also attend to the building and take care of the museum.

In December 1810, Secretary Smith wrote to John W. Eppes[8] concerning the pay for clerks employed in the Department of State. He said that Dr. Thornton, who superintends the issuing of patents for useful inventions, had said that there was too much work for one man to do in the timely fashion expected by patent applicants. Smith decided to postpone temporarily the replacement of a clerk who had resigned in order to have funds to hire an assistant in the Patent Office, and also to continue paying Dr. Thornton the $2,000 per year that he had been allowed the previous year. Smith said that this arrangement was temporary and would cease at the end of the year.

Smith said that William Thornton had a salary of $1,400 for superintending the issuing of patents for useful inventions, and carrying on all the correspon-

[8]See Robert Smith to John W. Eppes 12/21/1810 Federal Documentary Microfilm Edition No. 1/1.

dence incident thereto. George Lyon received a salary of $500 for performing those duties in the Patent Office that were assigned to him by Dr. Thornton.

In March 1811, William Thornton wrote a set of instructions for obtaining patents, which he published in the National Intelligencer newspaper.[9] These instructions were to enable many inventors to avoid long journeys to the seat of government. He was impressed with the idea that no nation on earth surpassed his countrymen in genius. Even the unlettered inhabitants of the forest had perfected inventions that would have done honor to Archimedes.

He suggested an examination of publications to ascertain if an invention were new and an inquiry of scientific experts to ascertain if it were practicable. The signatures of the President, the Secretary of State and the Attorney General on a patent should not be taken as evidence of novelty or utility. Before the organization of the patent system, the mysteries of a profession were withheld from the beginner, and many valuable secrets had been buried with their inventors. A patent is evidence that the inventor has disclosed his secret to the public and declares a right to prevent infringement, but only a jury can determine if others have infringed a protected right.

The instructions noted that the law formerly provided that only citizens could obtain patents, but a later law allowed for applicants who had resided two years in the United States but were not citizens.

Before the inventor presented his petition for a patent to the Secretary of State, he was required to pay into the Treasury of the United States $30 in notes of any of the state banks, for which he was furnished with duplicate receipts, one of which he delivered to the Secretary when presenting his petition. This money paid in full for the services of the office. The instructions gave the form of the petition and the requirements of the specification, which was required to be signed by the applicant before two witnesses. Thornton noted that the original papers were to be deposited in an office that will hand them down to posterity, so attention should be a matter of honor. The form of the oath or affirmation was given. The requirement for a good drawing, samples of ingredients, and a model of a machine, when appropriate, was set out. Thornton explained that: "Many will plead ignorance of drawings who cannot avoid the conviction of wheels and pinions," thus making a model necessary. The drawings ought not to exceed a quarto size, and if confined to octavo, it would be still better.

Because draftsmen were not readily available in many parts of the country, Thornton gave the names and addresses of eighteen of the best artists that he had noted around the country. Dr. Thornton noted that papers might all be mailed to the Secretary of State to render them free of postage, but when

[9]See *National Intelligencer* 3/12/1811.

models were sent, their freight to the Patent Office had to be prepaid. Before packing a model, the name of the inventor was be written on it, with the name of the machine, and the date. Sometimes, when models were received, it was difficult to know what application they related to.

The Congress, said Thornton, was so impressed with a high sense of the value of the inventions of our citizens that they purchased an "elegant and extensive building, wherein preparations are now being made for the accommodation of a very numerous collection" of models. This museum of the arts, he presumed, would stimulate the ingenious to send the models of their machines and inventions in a style that will rather honor than discredit our country. Thornton also gave instructions for securing copyrights of books, etc., by delivering a copy to the Secretary of State.

George Lyon was the first assistant ever hired for Dr. Thornton. He was appointed a clerk in June 1810, at a salary of $500 per year, even before the Patent Office moved to Blodgett's Hotel, although it is not certain that he was immediately assigned to patent duties. [10] For many months after his appointment, George Lyon hired a messenger for the Patent Office at his own expense at a cost of $4 per month. On April 1, 1811, the State Department stopped paying George Lyon his salary but did not fire him. In the early spring of 1812, Lyon was heard to complain that he had not been paid for nine months. Finally, Congress passed a private act in August 1813 for the relief of George Lyon to pay him his back salary.[11] George Lyon continued to work at the Patent Office almost until his death in 1817.[12]

Thomas Nicholson, carpenter, age 26, of Mecklenburg, Virginia, enlisted in the Army Ordnance Corps for five years[13] in August 1812. By December 1812, he was attached to the Patent Office, where he remained until May 1816. He seems to have been paid $250 per year[14] by the State Department. After returning to the Army, while still living in Washington, he obtained a patent of his own on an improved pump in April 1817.[15] Superintendent Thornton referred to Thomas Nicholson as his model-maker and messenger.[16]

[10]See George Lyon to Congress 4/3/1812 Federal Documentary Microfilm Edition No. 1/1.

[11]See 13th Congress 1st session Stat 1 Chap 47 Act of 8/2/1813.

[12]See [Edi59a].

[13]See National Archives M233 reel 9 p 199; Thomas Nicholson voucher 5/31/1816.

[14]See National Archives M235 reel 81 4/1/1813 *et seq* through M-235 reel 155 7/1/1816.

[15]See Thomas Nicholson patent of 4/29/1817.

[16]See *National Intelligencer* 9/7/1814.

Chapter 11

Dr. Thornton Saves Blodgett's Hotel from British

American independence had only recently been achieved when, in 1793, England began stopping American ships at sea, seizing any seamen who could not prove that they were not born in England and impressing them into service on British ships. England refused its subjects the right to become citizens of other nations. In this impressment operation, England seized many native-born American seamen, including at least one black seaman. Things escalated until war was declared in June 1812, which thus became the War of 1812. Much of the land war was fought around Lake Erie, and in April 1813, American troops burned the Canadian town of York (now called Toronto). At least partly in retaliation for the burning of York, the British sent an army of 3,500 under Major General Robert Ross to threaten Washington.

On August 20, 1814, British troops under the naval command of Admiral Sir George Cockburn landed at Benedict, Maryland, thirty five miles from Washington, and began marching toward Washington. It was obvious that the city could not be well defended with the few untrained militiamen from the area. All over Washington, government officials were attempting to pack up government papers and move them out of Washington and thus out of danger.

Dr. William Thornton still owned several pieces of land in the Washington area, including his farm of about one square mile in area, located in what is now downtown Bethesda, Maryland. This farm was eight miles from Blodgett's Hotel in the center of Washington. When Dr. Thornton learned of the approach of the British, he decided to move the Patent Office papers to this farm. Dr. Thornton, his clerk George Lyon, and his messenger and model maker Thomas Nicholson bought 402 feet of 3/4 inch wooden lathe and 100 feet of sheet iron for strapping,[1] then hired two men to assist them and made boxes for all of the papers and books of the Patent Office. [2] They filled the boxes with the papers, commandeered privately owned wagons and hired teamsters to haul the loaded boxes to the farm. They obtained wagons or teamster services or both from Burgess Willett, Charles Lemon, and Richard Fenwick.[3] The name of Richard Fenwick will appear again in Chapter 12 of this narrative.

Admiral Cockburn

General Robert Ross

Most of the time on August 22 and August 23 was spent packing and moving the Patent Office papers. But papers are relatively compact. Dr. Thornton had no way to move the patent models to safety. Models were as large as one cubic foot, and Thornton said there were then "hundreds" of them. [4] The patent models were left to their fate in Blodgett's Hotel. Dr.

[1]See George Moore voucher 8/22/1814 Federal Documentary Microfilm Edition No. 1/1; John C. Shindle voucher 9/28/1814 Federal Documentary Microfilm Edition No. 1/1.

[2]See Thomas Nicholson voucher 8/23/1814 Federal Documentary Microfilm Edition No. 1/1.

[3]See William Thornton to Burgess Willett 8/24/1814; Charles Lemon voucher 9/14/1814; Richard Fenwick voucher 10/3/1814, all Federal Documentary Microfilm Edition No. 1/1.

[4]See *National Intelligencer* 9/7/1814.

Thornton and his model-maker had been building a musical instrument which Dr. Thornton invented. It is unclear what this instrument was supposed to do, but apparently it imitated a wide variety of traditional instruments. The model, in whatever state of completion it had achieved at the time, was also left to its fate in Blodgett's Hotel.

Dr. Thornton wrote that on Tuesday, August 23, after securing all the public papers committed to his care and sending them to a place of perfect safety, he left his own papers unattended to and accompanied the Secretary of State Colonel James Monroe in reconnoitering the surrounding country, and returned home at midnight. While it seems odd that such an expedition should be led by the Secretary of State, Secretary of War John Armstrong was generally believed to be incompetent for the job, refused to admit that the British might invade Washington, and resigned soon after the British invasion of Washington. James Monroe then became both Secretary of State and Secretary of War. The rapid departure, shall we say rout, of the local militia from the engagement with the British at Bladensburg, Maryland, on Wednesday, August 24, was derided by the local residents as the Bladensburg Races. Only Navy and Marine troops under Commodore Joshua Barney distinguished themselves.

On August 24, Thornton moved his family with the retreating army from the city of Washington to Georgetown. That night, he observed the fires set by the British burning the public buildings of Washington. While having breakfast in Georgetown on Thursday, August 25, he was told that the British were preparing to burn the War Office and Blodgett's Hotel. He rode over to Washington to try to save his musical instrument. He arrived in time to see the War Office burned. Along with Thomas Nicholson, George Lyon and two others, Thornton approached Major Waters, who was awaiting the command of Colonel Timothy Jones to burn Blodgett's Hotel. There were 150 soldiers[5] marching toward Blodgett's Hotel to do the job. Thornton asked Major Waters if he could remove his musical instrument and was told that it was not the British intention to destroy any private property and that he could remove any private property from the building.

Here, Dr. Thornton hit his stride. The generation in the Patent Office after Thornton's death would magnify this incident until it bore only the faintest resemblance to the incident as it happened. Thornton himself reported in a newspaper account that he told Major Waters there was nothing but private property in the building of any consequence and that he should remove any public property that he objected to from the building and burn it in the street.

[5]See William Thornton to John W. Taylor 2/14/1821 Federal Documentary Microfilm Edition No. 1/2.

He said that the building contained hundreds of models of the arts and that it would be impossible to remove them. He said that the models were useful to all of mankind, not just to Americans, and that anyone who burned them would be condemned by future generations as were the Turks who burned the Library at Alexandria.

Dr. Thornton, in his impassioned speech given despite a very noticeable tendency to stutter or stammer, confused Major Waters sufficiently that the major asked Thornton to accompany him and present his request to Colonel Jones. Thornton went with the major and found Colonel Jones supervising the destruction of Joseph Gales' printing establishment. Mr. Gales' newspaper, the National Intelligencer, had been printing articles which were derogatory to Admiral Cockburn, and the admiral had given orders that special care should be taken to destroy all type for the letter C so that Gales could not take his name in vain again. (Mr. Gales borrowed type and printed his newspaper again in one week. The admiral was mentioned.)

After Dr. Thornton presented his case, Colonel Jones ordered the men away and spared Blodgett's Hotel. At least, that was Thornton's version. Washington's Mayor James H. Blake, apparently feeling libeled by Thornton's newspaper account of their respective actions during the invasion, wrote in his own newspaper account that he would give Thornton credit for saving the building, notwithstanding there were many who think the building was saved by the storm which happened later that day. Indeed, a hurricane visited Washington City that afternoon. The oldest residents said they could not recall a storm so intense. The storm extinguished those fires that were still burning and prompted the immediate withdrawal of the British from the city. It also blew part of the roof off Blodgett's Hotel.[6]

Sometime later, while he was a prisoner on a ship in Baltimore Harbor, Dr. William B. Beanes (1749–1828) of Upper Marlboro, Maryland, a colleague and friend of Dr. Thornton, overheard British soldiers discussing the actions of Thornton in saving Blodgett's Hotel.[7] It was while negotiating to secure the release of Dr. Beanes that Francis Scott Key, a Washington lawyer, was forced to spend the night aboard this same ship, witnessing the naval bombardment of Fort McHenry. On the next morning, before Key and Beanes left the ship, Key noticed with pride that the American flag still flew over Fort McHenry, and in a burst of inspiration that morning, September 14, 1814, Key wrote the words to the Star-Spangled Banner. The words were immediately set to

[6]See *National Intelligencer* 9/9/1814 and [Gle47].

[7]See William Thornton to William B. Beanes 3/31/1824 Federal Documentary Microfilm Edition No. 1/1.

the music of an old song, and the song was performed in Baltimore before the British had finished leaving the Chesapeake Bay.

As the British left Washington City, Congress was due to arrive for a session in two weeks. The Capitol had been destroyed, and Congress needed a place to meet. The only undamaged government building left in Washington City was Blodgett's Hotel. The Patent Office and most of the Post Office were told to seek other quarters,[8] clear the building, and make it available as a meeting place for Congress. This they did, and for over a year the Congress of the United States met in the Patent Office[9] and Post Office quarters.

There was much agitation to move the seat of government to some large city where there was sufficient population to protect it from invasion, and, incidentally, sufficient accommodations so that Congressmen did not feel like they were camping out during legislative sessions. It is clear that if Blodgett's Hotel had not been available to provide sufficient space for Congress to meet in Washington City, there would have been enough votes to move the seat of government. However, those who wished to move the government failed, barely, to obtain enough votes while in their new temporary quarters.

It could be said that on the morning of August 25, 1814, Dr. William Thornton set forth to Blodgett's Hotel to save a musical instrument, and in the course of the day saved first Blodgett's Hotel itself and then the seat of government. Congress continued to meet in Blodgett's Hotel until a group of local citizens invested their own money to build the Old Brick Capitol on the present site of the Supreme Court Building.

During their exile from Blodgett's Hotel, the staff of the Patent Office conducted business from the "Den" in a private house.[10] It has been suggested that the Den was the home of George Lyon.[11] During this period, rent of $120 per year was being paid to William Cocking[12] for the house occupied by the Patent Office. The Patent Office papers were returned from William Thornton's farm to the Den about September 12, 1814.[13] By December 1815, the Patent Office was preparing to move back to Blodgett's Hotel, and Thomas Nicholson was paid for tearing out the remains of the old lobby to the Congress

[8]See James Monroe to William Thornton 9/3/1814 Federal Documentary Microfilm Edition No. 1/1.

[9]See [Edi59a] and William Thornton to John Quincy Adams 8/10/1824 Federal Documentary Microfilm Edition No. 1/3.

[10]See Thomas Nicholson voucher 3/3/1815 Federal Documentary Microfilm Edition No. 1/1.

[11]See [Edi59a, p. 263].

[12]See William Cocking voucher 1/12/1816 Federal Documentary Microfilm Edition No. 1/1.

[13]See James W. Johnston voucher 3/3/1815 Federal Documentary Microfilm Edition No. 1/1.

Hall and for doing other work to prepare the building for Patent Office occupancy.[14] In early January 1816, the boxes, cases, books, firewood and coal of the Patent Office were moved back to Blodgett's Hotel.

[14]See Thomas Nicholson voucher 12/18/1815 Federal Documentary Microfilm Edition No. 1/1.

Chapter 12

The Rest of Dr. Thornton's Tenure

John Smith

Most of the inventors who thread through our history are relatively unknown, and sometimes the deeds we know of do not connect with a name which has come down to us. We know most of the inventors of the dramatic inventions, such as Samuel Morse, Thomas Edison, Eli Whitney and Charles Goodyear. But the man who added a cam to the bottle-making machine which pushed down on a tab to expel the bottle faster than could be done by gravity, thereby increasing its speed in making bottles by twenty percent, falls into the large group of inventors of useful machines which improve efficiency and save money but which do not dramatically affect the life of anyone. The generic inventor herein is named *John Smith*, with all apologies to all those John Smiths in the world who are not *John Smith*.

John Smith was born in 1776, 1777, 1778, 1779, 1780, and every subsequent year. He seems to be bred more easily in America than in most other places, although he frequently arrives here from foreign shores named Johann Schmidt, Giovanni Fabrizio, or Li Wu. Here we arrive at, certainly not the first, but the first-mentioned *John Smith*. *John Smith* is so named, not because that is his name, but because we know of no other name for him.

Our first *John Smith* was a black man, who lived on Hempstead Plain, Long Island, New York. Of his birth, we know nothing. Of his death, we know that it occurred about 1821. Of his life, we know only that he invented the horse-drawn rake, an implement which has allowed succeeding generations of farmers to gather in their hay crops without raking the hay by hand. His invention was introduced by Michael Newbold in 1812 into Pennsylvania, where the first rake was destroyed by an angry farmhand who feared its effect on his wages.[1] But the rake itself succeeded beyond belief, spreading around the world and inspiring many improvements. *John Smith* did not patent his invention, probably did not profit from it, and passed into near anonymity.

In a forewarning of things to come, Postmaster General Return Jonathan Meigs wrote a letter to Dr. Thornton on December 11, 1816,[2] complaining of the danger of fire in the building they jointly occupied. The Patent Office had established a workshop in which models were painted, and the use of paint in the same room with a fireplace was considered dangerous.

William Elliot (ca 1773–December 30, 1837) was from Carlisle, Cumberland, England. William Faux[3] wrote that Mr. Elliot had been a neighbor of Archdeacon William Paley, and Paley began his career in Carlisle. William Elliot began publishing the Washington City Gazette in January 1814, and its publication continued with gaps until November 1817, under the control of Jonathan Elliot, probably his brother, also from Carlisle. William Elliot was also a surveyor and an astronomer. Mr. Elliot sold stationery and printed patent-cover sheets to the Patent Office.[4] On July 1, 1816, William Elliot was hired as a clerk in the Patent Office, probably replacing George Lyon. This position did little to limit his other professional activities. William Elliot remained a clerk in the Patent Office until his resignation on May 10, 1829.[5]

Richard Fenwick (ca 1759–March 26, 1829) was from St. Marys County, Maryland, where he was a miller from 1804 to 1812. Apparently, he had moved to Washington City by the latter part of this period.[6] He was employed as a porter by the Bank of the Metropolis. Among his children were Richard Washington Fenwick (ca 1798–Oct 29, 1832),[7] Benjamin Fenwick (ca 1800 - December 8, 1820),[8] and Robert Welsh Fenwick (ca 1804–January 19, 1845).[9] The Richard Fenwick who drove the wagon to transfer Patent Office papers to Dr. Thornton's farm in 1814 may have been the father or the son. Looking

[1]See [[Jo39].

[2]See Return Jonathan Meigs to William Thornton 12/11/1816 Federal Documentary Microfilm Edition No. 1/1.

[3]See [Fau23].

[4]See William Elliot voucher 10/13/1815 Federal Documentary Microfilm Edition No. 1/1.

[5]See papers of Pasquale Joseph and Bianca Morse Federico, Smithsonian National Museum of American History Library and *National Intelligencer* 1/1/1838.

[6]See [Mar79], [Del22], and National Intelligencer 3/27/1829.

[7]See *National Intelligencer* 11/2/1832.

[8]See *National Intelligencer* 12/9/1820.

[9]See *National Intelligencer* 1/21/1845.

back, we can only say the driver could have been 16 or 55, and we cannot choose between them with certainty.

In December 1814, Benjamin Fenwick was hired to cut and pile firewood for the Patent Office.[10] Dr. Thornton later hired Benjamin as a messenger to assist him, paying him $6 per month from his personal funds.[11] Benjamin Fenwick was listed in the register of government employees[12] as of September 1816, at $72 for the year, perhaps meaning that he started work at $250 per year in mid-June 1816, or perhaps that he was actually earning $6 per month. In the register for September 1817,[13] he was earning $250 per year. The first indication that Dr. Thornton had hired him at personal expense was in a letter of September 19, 1817,[14] indicating that he had been hired at least as early as April 1817. The sum of seventy-five cents was spent in October 1817 for making a green bag for Ben to carry letters and other papers in for the Patent Office. He seems to have kept this job until replaced by his brother Robert Welsh Fenwick in April 1819.[15] Benjamin died on December 8, 1820, after an illness of twelve weeks.

John Smith, or the same man going by another name, made inventions after William Elliot came to the Patent Office in 1816 and before William Thornton's death in 1828. He walked to the Patent Office from his home in Wheeling, Virginia, or perhaps it was Binghamton, New York, or maybe New Bern, North Carolina, carrying a model of his improved saddle on his back because he could not afford a horse to ride.[16] Or perhaps I am mistaken and the invention was an improved plow or an apple-peeler. It is not important. He was sure he would be made rich when the public learned of his invention. He had no money to pay for a patent application. Perhaps he could sell a license for a county or even a whole state to get money to pay for the patent. If not, Dr. Thornton did not believe the patent system was set up to make money for the government, but it did nevertheless. After Dr. Thornton's death, William Elliot testified that Thornton sometimes issued patents to impoverished inventors with worthwhile

[10]See Benjamin Fenwick voucher 11/26/1814 Federal Documentary Microfilm Edition No. 1/1.

[11]See William Thornton to Moses Young 9/29/1817 Federal Documentary Microfilm Edition No. 1/2.

[12]See Register of the Officers and Agents of the United States as of September 30, 1816.

[13]See Register of the Officers and Agents of the United States as of September 30, 1817.

[14]See William Thornton to Moses Young 9/19/1817 Federal Documentary Microfilm Edition No. 1/2.

[15]See William Elliot voucher 10/6/1817 Federal Documentary Microfilm Edition No. 1/2; William Elliot to Daniel Brent 3/31/1819 Federal Documentary Microfilm Edition No. 1/2; National Archives M235 reel 155 account 32,472.

[16]See [Ell37].

inventions without the payment of any fee.[17] Apparently, the audit system was so ineffective that no one ever compared the number of patents issued with the amount of money paid into the Treasury for patents.

The Act of February 21, 1793, required a patent applicant, before petitioning the Secretary of State for a patent, to pay into the Treasury the sum of thirty dollars, to take duplicate receipts for that money, and to send one of the receipts to the Secretary of State to pay all of the expenses of the patent. This was in the days before the public wrote checks, and when even the most educated and successful men sent cash through the mail. There were Treasury offices in most major cities, so receipts could be obtained locally and sent to Washington without risk of loss. However, once the money had been paid into the Treasury, if the applicant decided that he no longer wished to obtain a patent, he required a private act of Congress to get his money refunded.

Dr. Thornton's practice differed somewhat from the law. The law did not allow the Patent Office to refuse patents for lack of novelty. There was no requirement that the applicant claim specific features of his device as the novel part of his invention. But Thornton still examined applications, comparing them with prior patents and with the large library of technology books kept for the purpose. When he found a good reference to anticipate an invention, he would write to the applicant, explaining the reference and pointing out how limited the patent on the invention would be if it were allowed to issue. Then, if the applicant still insisted on his patent, Dr. Thornton would write across the back of the issued patent the identity of the known references. However, if the applicant had already paid in full for his patent, then the applicant had no financial incentive to refuse a patent of doubtful validity. Thornton would accept cash sent to him through the mail, and, if in his opinion the application warranted a patent, he would deposit the money with the Treasury and issue the patent. Until this point, Dr. Thornton had kept the money deposited in a local bank account.[18] If Thornton convinced the applicant to withdraw his application, he could return the cash, which had never been deposited in the Treasury, without the necessity of a private act passed by Congress to allow its return.

With all this cash floating around, there was a risk of loss, either in the mail, or through theft or embezzlement, or through bank failure. In 1816, Dr. Thornton had \$120 in application fees stolen from his desk.[19] In 1817,

[17]See William Elliot to John D. Craig 1/16/1830 in Doc 38, 31st Congressional 1st Session, House of Representatives.

[18]See William Thornton to James Monroe 2/19/1816 Federal Documentary Microfilm Edition No. 1/1.

[19]See William Thornton to James Monroe 2/19/1816 Federal Documentary Microfilm Edition No. 1/1.

Peter Arrell Browne (1782–1860), a native of Philadelphia, wrote to Acting Secretary of State Richard Rush, complaining that he had sent $30 through the mail to Thornton, which Dr. Thornton denied having received.[20] In 1834, some years after the death of Thornton, a later Superintendent was instructed to pay immediately into the U.S. Treasury all cash received by the Patent Office for payment of application fees, thus requiring a private act of Congress to return any of it to applicants.[21]

George Escol Sellers

George Escol Sellers (1809–1897), a grandson of Nathan Sellers and of Charles Willson Peale, lived in the midst of much of the nation's technological history and had a very retentive memory.[22] He wrote[23] of a trip he made by wagon with his father and Oliver Evans in 1818 when he was nine years old. Evans had spoken of difficulties facing inventive mechanics with no published records of what had preceded them. They had to start at the very foundations of mechanics and repeat work previously done by others, and often abandoned as unsuccessful, because they did not know what had been done before. Evans wanted a Mechanical Bureau set up to collect and publish all new inventions, together with reliable treatises on known areas of mechanics. He wanted a school set up to teach mechanical drawing. There was then no school in the country where a young man could go to learn any kind of engineering other than the military engineering taught at the United States Military Academy. The term civil engineering was first defined in 1818 in London, and encompassed all non-military engineering. The first school of civil engineering in America was opened in 1824. Sellers noted that Evans never lost an opportunity to impress his listeners with the feasibility of using steam, not only to navigate rivers, but also to cross oceans and continents.

Oliver Evans (1755–1819) was a man of genius, ahead of his time and ahead of his generation. He invented the steam engine from basic principles at age

[20]See Peter Arrell Browne to Richard Rush 4/8/1817 Federal Documentary Microfilm Edition No. 1/2.

[21]See John Forsyth to John D. Craig 10/1/1834 National Archives M40 reel 25 pp 58–59.

[22]See [Sel65].

[23]See [Sel84].

eighteen in 1773,[24] notwithstanding that, unknown to him, it had previously been invented by others in England and perhaps elsewhere. He published a book on the mechanics of milling[25] which was continuously in print in this country until after the Civil War, for use by millers, and is still in print for use by historians. He got the third patent ever issued in this country; it was for an improved system of milling. He was informed by the judge of a United States court that a patent right for a useful improvement was an infringement of the public right. Evans wrote that when he heard this judgement, he destroyed all of his papers relating to invention[26] and went to work for his private interests and became wealthy, using the same talents which, when employed for the public interests, had kept him in poverty.[27]

Orukter Amphibolis

In 1804, he built what was probably the first mechanically driven land vehicle in the United States. He was hired to build a steam-driven dredging machine for cleaning the Philadelphia docks. He built his machine as a boat, called the Orukter Amphibolis, or amphibious digger, at his shop about a mile and a half from the docks where its use was intended. As a demonstration that steam-driven land vehicles were possible, he put wheels on his boat and drove

[24]See [Eva17].
[25]See [EE07].
[26]See [Eva14a].
[27]See [Eva14a].

it under steam power down Market Street, in what was then the largest city in the country, then several times around Center Square and up to the banks of the Schuylkill River. He then attached a paddle wheel to the stern of the boat and powered it up the river a distance of 16 miles and back, to the amazement of thousands of spectators.[28]

But Oliver Evans' most astounding feat was not in what he did, but in indicating what he might have done if the patent laws of the day had been good enough to make it possible to try. He wrote a prophecy[29] in 1814 that "The time will come when people will travel in stages moved by steam engines, from one city to another, almost as fast as birds fly, fifteen or twenty miles in an hour. Passing through the air with such velocity, changing the scene in such rapid succession, will be the most exhilarating, delightful exercise. A carriage will set out from Washington in the morning, the passengers will breakfast in Baltimore, dine in Philadelphia, and sup at New York, the same day. To accomplish this, two sets of rail ways will be laid, so nearly level as not in any place to deviate more than two degrees from a horizontal line, made of wood or iron, or smooth paths of broken stone or gravel, with a rail to guide the carriages, so that they may pass each other in different directions, and travel by night as well as by day; and the passengers will sleep in these stages as comfortably as they now do in steam boats."

No wonder George Escol Sellers, at the age of 77, remembered clearly the conversation he had with Oliver Evans when Sellers was nine. Nowadays, we sometimes package such conversations a little differently. They are now sold as science fiction, including what might someday be science. The first steam-driven passenger train came into existence 11 years later in 1825 in England.

Dr. Thornton continued in his practice of attempting to insinuate himself into association with inventors in order to benefit financially from inventions which were brought before him. The most notorious case was his revision of the application of John H. Hall in 1811 to make himself a joint inventor,[30] but there were others, including Jacob Cist, Robert Fulton, and Abel Brewster. Mr. Brewster wrote:[31]

> One time I was at Washington, application was made for a patent (for what I believe was a Pipe Boiler). I understood Doctor Thornton claimed the invention; but when he was asked to produce for me evidence of his previous knowledge of it, he replied that he

[28]See [Lat72].

[29]See [Eva14b].

[30]See [Smi77, Chapter 7] and [Pre85].

[31]See Abel Brewster to John Quincy Adams 1/20/1818 Federal Documentary Microfilm Edition No. 1/2.

> could describe it. To which the applicant remarked that, very likely he could, since the specification and drawings had been shown to him... A story which Judge Livingston is said to relate, as a burlesque upon some of Doctor Thornton's transactions—A man applied for a patent for making boards out of sawdust—The Doctor said it was nothing new—he had known it long before; But said the applicant, my invention is for making oak boards out of pine sawdust!

Dr. Thornton had a long controversy with Michael Withers,[32] alleged inventor of a winged gudgeon, whatever that might be. There was a model of Mr. Withers' winged gudgeon in the Model Room of the Patent Office, sometimes also known as Dr. Thornton's Toy Shop. On this model, William Thornton had posted a notice[33] in his own handwriting reading: "This model is an imposition, having been deposited instead of the original which was taken from the office, by a person unknown, but strongly suspected. [signed] William Thornton March 24th 1819" Most probably the person who was suspected by Dr. Thornton was Michael Withers.

John Quincy Adams

Possibly the first office ever opened in Washington for full-time practice as a patent agent was that of William Blagrove, who advertised in 1819 that he was opening an office "at the Seat of Government" to secure patents and copyrights for inventors and authors.[34] Oddly enough, no further address was given or perhaps considered necessary. Later the same year, he wrote a proposal[35] for publishing a register of inventions and patent laws, but if it was ever published, there is no record of it.

Secretary of State John Quincy Adams wrote in his diary in April 1819 that he had received a letter from an inventor which caused him to visit the Patent Office, apparently for the first time.[36] Thornton gave him a tour of the Model Room.

[32]See See Michael Withers to Henry Clay 3/22/1827 and accompanying papers, Federal Documentary Microfilm Edition No. 1/4.

[33]See William Thornton statement 3/24/1819 Federal Documentary Microfilm Edition No. 1/2.

[34]See *National Intelligencer* 4/19/1819.

[35]See William Blagrove proposal 11/22/1819 Federal Documentary Microfilm Edition No. 1/2.

[36]See diary of John Quincy Adams, 4/26/1819.

He wrote:

> I thought how useful and profitable an occupation it might be for a young man with a competent fortune, and having no other necessary pursuit in life, to take up this collection of models, to examine and make himself thoroughly master of the principles, and of the peculiar invention, or new idea, with its application, in each of them, then to classify them, to mark all those, if any there be, which contain a new principle, to examine the differences in the modifications of the same principle, then to observe and compare them with reference to their ultimate results—what effect they produced upon the enjoyments or conveniences of human life, distinguishing those of which the ends are comprehensive and important from those which terminate in trifles ... Would it not be worth while, among the public institutions of the nation, to have a school for the education of a certain number of civil engineers?

Gilbert Brewster carried out part of Secretary Adams' mandate in 1823. As Dr. Thornton later reported:[37]

> Mr. Gilbert Brewster, a very ingenious artist, from Connecticut, came to the Patent Office about the middle of October, 1823, and requested permission to examine the models. I informed him they were deposited for public inspection, and he was at liberty to see and examine them as often as he pleased. Instead of spending a few hours, he visited them daily for about six weeks; then thanked me for the gratification he had enjoyed, declaring them worth millions of dollars, or that they were of incalculable value to a real mechanician. He said he saw movements and combinations of which he had before no idea, and that he was now enabled so to improve the machinery for spinning wool, as to reduce the price of ginning from eight cents to one cent per pound. He went away, and returned in about three months, with two models, declaring, on his return, that he had perfected what he had contemplated, and that he could then spin wool at a lower price than the English, who could not effect it for less than four cents per pound. I issued three patents for his machines, and a gentleman who accompanied him from New York, and who had engaged to buy these machines for a manufacturing company in Connecticut, laid him down ten thousand dollars in my presence."

[37]See William Thornton to President of Convention for Promotion of Manufactures 7/23/1827 Federal Documentary Microfilm Edition No. 1/4.

In October 1819 came the beginning of something brought about by serendipity that was to have beneficial effects on the patent system and the nation to the present day. Jonas Keller (ca 1768–1830), an immigrant from Switzerland, was a skilled mechanic who had come to Washington to start a stocking factory. When that failed, he worked where he could, but the returns were inadequate. A number of his acquaintances wrote to Secretary Adams[38] to recommend his employment in the Patent Office to mend and arrange the deteriorating invention models. When Dr. Thornton was asked what he thought about such an appointment, he replied that it would be a good idea but that there was no appropriation to pay for it.[39] He said when Congress was to meet in Blodgett's Hotel, the models were removed from the Model Room, then called the Museum of the Arts, into the garrets above, and then later back into the Model Room. All of this moving and piling of the models had caused many of them to be broken and injured and in much need of repair. Keller was not appointed mender of models immediately, but in November 1820 he was paid for fitting a screw to the Great Seal of the United States.[40] In August 1822, Superintendent Thornton wrote to Secretary Adams,[41] explaining the deteriorating condition of the patent models and recommending that Jonas Keller be hired to repair them. Daniel Brent, the Chief Clerk of the State Department, replied that Keller could be employed at two dollars a day until Congress met to consider the matter.[42] In an annotation to Jonas Keller's first voucher for his work,[43] Dr. Thornton noted that Mr. Keller had worked very attentively and had not only mended many of the models, but had restored a great many by new work. In January 1823, William Thornton informed Secretary Adams that Jonas Keller not only repaired the models but also exhibited them to visitors including members of Congress.[44] Not only that, "but his son, a very ingenious boy, capable of doing good work, has also been constantly employed in the office assisting him, and for his services he has

[38]See J.B. Colvin et al to John Quincy Adams 10/2/1819 Federal Documentary Microfilm Edition No. 1/2.

[39]See William Thornton to John Quincy Adams 11/1/1819 Federal Documentary Microfilm Edition No. 1/2.

[40]See 5th Auditor account 962 11/3/1820 Federal Documentary Microfilm Edition No. 1/2.

[41]See William Thornton to John Quincy Adams 8/22/1822 Federal Documentary Microfilm Edition No. 1/3.

[42]See Daniel Brent to William Thornton 8/26/1822 Federal Documentary Microfilm Edition No. 1/3.

[43]See Jonas Keller voucher 10/4/1822 Federal Documentary Microfilm Edition No. 1/3.

[44]See William Thornton to John Quincy Adams 1/4/1823 Federal Documentary Microfilm Edition No. 1/3.

not made any charge whatever." In March 1823, in another letter to Adams,[45] Thornton wrote: "Mr. Keller employed, not only his own time very faithfully and honestly, but also the time of his son, who is likewise very ingenious, and who was engaged daily, not only in assisting his father in his various duties, but also in exhibiting and explaining to the daily visitors the machines, and was more useful than I could have supposed a boy of his years could have been." Mr. Keller's son was Charles Michael Keller (ca 1809–1874), who at this time was about thirteen years old. As will be seen later, it is the early and prolonged interest of Charles M. Keller in the Patent Office that provides that serendipity. Much more will be heard from this boy as a man.

In July 1820, Nathan Sellers of the firm of Nathan and David Sellers, he who some 44 years earlier had been excused from the Revolutionary Army to make wire screens for paper-making, he who was grandfather to George Escol Sellers, applied for a patent for a joint invention of him and his late brother David.[46] The invention was for a rolling wire screen to clean grains, particularly rice, as well as sand, shot, and other items. This was soon issued to Nathan Sellers, on his own behalf and as administrator for his late brother.[47]

Peter Arrell Browne

In an aside to a controversy between William Thornton and Joseph Hawkins, William Elliot wrote to Mr. Hawkins in 1823,[48] indicating that it took about three weeks from receipt of an application for patent until the specification would be copied in due course and be ready for signature. This was considered too much time by the applicants of the day.

In August 1823, William Parker Elliot (ca 1807–1854), one of the sons of William Elliot, submitted a voucher[49] for his services in the Patent Office during the last quarter. He later indicated that he had been doing copying work in the Patent Office since 1817.

[45]See William Thornton to John Quincy Adams 3/7/1823 Federal Documentary Microfilm Edition No. 1/2.

[46]See Nathan and David Sellers to William Thornton 7/20/1820 Federal Documentary Microfilm Edition No. 1/2.

[47]See William Thornton to Nathan Sellers 7/29/1820 Federal Documentary Microfilm Edition No. 1/2.

[48]See William Elliot to Joseph Hawkins 3/27/1823 Federal Documentary Microfilm Edition No. 1/3.

[49]See William Parker Elliot voucher 8/20/1823 Federal Documentary Microfilm Edition No. 1/3.

By 1824, a mechanic's institute had been formed in Philadelphia along the lines proposed by Oliver Evans to George Escol Sellers and others before his death in 1819. This was the Franklin Institute of the State of Pennsylvania for the Promotion of the Mechanic Arts, usually called the Franklin Institute. In March 1824, on its behalf, Peter A. Browne, corresponding secretary, wrote a letter to William Thornton explaining its goals and asking for favorable terms for obtaining informal copies of patents being issued for publication in its journal.[50] Despite the Attorney General's decision in 1809[51] that copies of issued patents should be available upon request, Thornton replied that he would resist providing copies of unexpired patents except where required by a law suit or when authorized by the patentee.[52] He noted that for copies which were made by the Patent Office, the statute required payment of 20 cents per 100 words, although others outside the office might be engaged to copy them at a lower price. In early 1825, Mr. Browne wrote to Thornton again,[53] asking if he would still refuse to provide copies of specifications of patents until their expiration. When he did not receive a reply within nine days, he wrote a second, rather insulting letter to Thornton demanding a reply, which set the tone for their later dealings, if it had not already been set by his 1817 correspondence.[54] Thornton noted in a reply[55] that he received the original and the second letter on the same day, and that their delay was occasioned by the fact that Mr. Browne had not prepaid the postage, thereby requiring their delivery to the Secretary of State, who alone had the franking privilege. He objected to being insulted and replied that his earlier letter still expressed his intentions. Browne apologized vaguely,[56] indicating that when he wrote his letter he had every reason to believe Thornton was being incivil to him in not answering his letter promptly, but following the explanation, he regretted having used the offensive expressions. He did not apologize for assuming that offense was intended on such slender evidence. Dr. Thornton was exasperating

[50]See Peter A. Browne to William Thornton 3/6/1824 Federal Documentary Microfilm Edition No. 1/3.

[51]See John Stevens to William Thornton 7/28/1809 Federal Documentary Microfilm Edition No. 1/1.

[52]See William Thornton to Peter A. Browne 3/10/1824 Federal Documentary Microfilm Edition No. 1/3.

[53]See Peter A. Brown to William Thornton 1/7/1825 Federal Documentary Microfilm Edition No. 1/3.

[54]See Peter A. Browne to Richard Rush 4/8/1817 Federal Documentary Microfilm Edition No. 1/2.

[55]See William Thornton to Peter A. Browne 1/29/1825 Federal Documentary Microfilm Edition No. 1/3.

[56]See Peter A. Browne to William Thornton 1/31/1825 Federal Documentary Microfilm Edition No. 1/3.

to deal with, and Peter Browne was too petulant to succeed in dealing with him.

Mr. Browne then wrote to Secretary Adams[57] before receipt of Thornton's reply, asking that the copies be furnished to the Franklin Institute and without charge. He waited over two weeks before sending a follow-up letter to Mr. Adams.[58] Daniel Brent, Chief Clerk of the State Department, soon wrote to Mr. Browne, acknowledging receipt of the letters "by direction of the Secretary" and added that they would be attended to as soon as the Secretary found time from his more important duties.[59] Within less than a month, and soon after the inauguration of John Quincy Adams as President, Mr. Browne wrote a letter to the new Secretary of State, Henry Clay,[60] filing charges that the persons having the care of the Patent Office were grossly ignorant of the law relating to the office, were grossly partial in the exercise of their official duties, and were grossly inconsistent in their official conduct. Mr. Browne wrote two weeks later to Mr. Adams as President,[61] repeating his charges, including the charge that he had written to Dr. Thornton in January and Thornton would not even answer his letters. This was done after his half-hearted apology to Thornton for making the same charge. Then he indicated his belief that since Mr. Clay had not responded to his charges, he had not been permitted to see them. Secretary Clay replied[62] that he had seen the letters, but that pressure of official business had prevented him from personally answering them. He said that if Mr. Browne had proof of his charges, he should transmit it.

Most of these charges have little in them of lasting interest. Mr. Browne filed his full set of charges in four days,[63] and in another four days William Thornton filed his reply[64] characterizing the charges, probably correctly, as malevolent and denying all the charges.

---

[57]See Peter A. Browne to John Quincy Adams 1/29/1825 Federal Documentary Microfilm Edition No. 1/3.

[58]See Peter A. Browne to John Quincy Adams 2/14/1825 Federal Documentary Microfilm Edition No. 1/3.

[59]See Daniel Brent to Peter A. Browne 3/19/1825 Federal Documentary Microfilm Edition No. 1/3.

[60]See Peter A. Browne to Henry Clay 3/7/1825 Federal Documentary Microfilm Edition No. 1/3.

[61]See Peter A. Browne to John Quincy Adams 3/23/1825 Federal Documentary Microfilm Edition No. 1/3.

[62]See Henry Clay to Peter A. Browne 3/25/1825 Federal Documentary Microfilm Edition No. 1/3.

[63]See Peter A. Browne to Henry Clay 3/29/1825 Federal Documentary Microfilm Edition No. 1/3.

[64]See William Thornton to Henry Clay 4/2/1825 Federal Documentary Microfilm Edition No. 1/3.

Henry Clay

Secretary Clay requested an opinion from Attorney General William Wirt[65] on whether the law required the Patent Office to furnish a copy of an unexpired patent to a member of the public who demanded it, and whether such copies could be furnished by the office at a cost of less than 20 cents per 100 words. Mr. Wirt replied that the Patent Office was not required to furnish copies to anyone who demanded them, but that the sound discretion of the State Department could be used to determine when they ought to be furnished. He also wrote that it would not be legal for copies to be issued from the Patent Office at less than the rate required by the statute, which was 20 cents per 100 words.[66]

Apparently, no reply was provided to Peter Browne concerning the charges of ignorance, partiality and inconsistency. Mr. Clay did provide a copy of Mr. Wirt's opinion to Peter Browne.[67] Mr. Browne wrote with unusual propriety and restraint to Secretary Clay,[68] including a copy of an opinion by eminent Philadelphia lawyers Horace Binney and John Sergeant.[69] Upon consideration of the Binney and Sergeant opinion, Henry Clay decided in September 1825 to exercise his discretion and allow the Franklin Institute to obtain copies of unexpired patents.[70]

The present generation in and around the Patent Office could not imagine a reason why a person about to manufacture an article in the United States would not be allowed to see any patent in existence, expired or unexpired, to determine what the law said his rights were. Failure of the Patent Office to allow copies of unexpired patents so that a manufacturer might avoid violation

[65]See Henry Clay to William Wirt 4/15/1825 Federal Documentary Microfilm Edition No. 1/3.

[66]See William Wirt to Henry Clay 4/16/1825 Federal Documentary Microfilm Edition No. 1/3.

[67]See Henry Clay to Peter A. Browne 4/20/1825 Federal Documentary Microfilm Edition No. 1/3.

[68]See Peter A. Browne to Henry Clay 5/11/1825 Federal Documentary Microfilm Edition No. 1/3.

[69]See Horace Binney and John Sergeant opinion 5/10/1825 Federal Documentary Microfilm Edition No. 1/3.

[70]See Henry Clay to Peter A. Browne 9/12/1825 Federal Documentary Microfilm Edition No. 1/3.

of these patents is nearly equivalent to failure of the government to allow an individual to have a copy of the statutes so that he might avoid violation of these statutes. We owe the availability of these patents, at least in part, to the persistence, if not to the petulance, of Peter Arrell Browne. In early 1826, Dr. Thomas P. Jones began the publication of the Journal of the Franklin Institute, which gave abstracts or sometimes full text of recently issued patents, together with Dr. Jones' penetrating analysis of the value of the new inventions or lack thereof. For the first time, information on recently issued patents which had previously been available only in the Model Room of the Patent Office was available for the price of a subscription anywhere in the country.

In July 1825, the Patent Office got its own horse, having previously been dependent upon use of the State Department horse. Robert Welsh Fenwick was paid $10 per month for keeping the Patent Office pony, beginning August 1, 1825.[71] Later, in April 1826, an iron gray horse was purchased for use by the Patent Office at a cost of seventy dollars.[72] In June 1827, Robert Welsh Fenwick was paid $2.50 for his expenses in trying to recover the Patent Office horse, which had been stolen.[73] In August, he was paid $5.00 for his expenses in recovering the horse from W. Maul.[74] The Patent Office pony was thus stolen and recovered three months later. Alexander M. McIntire was appointed as a temporary clerk in March 1827 to assist in copying issued patents into record books as had long been required by law, but which had not been done for lack of clerical force.[75] He had earlier been employed as a copyist on a daily basis as early as the spring of 1825.[76] He was to feature in some later controversy in the Patent Office.

In early March 1828, John McLean of the Post Office Department wrote to Secretary of State Henry Clay, stating that the Post Office Department had grown so much that it needed Blodgett's Hotel for its exclusive use, and suggesting that the Patent Office be moved to the Old Brick Capitol, which had been erected for the temporary use of Congress as it moved out of Blodgett's

[71]See William Browne account, 5th auditor account 1686 9/30/1825 Federal Documentary Microfilm Edition No. 1/3.

[72]See C.W. Boteler voucher 4/19/1826, 5th auditor acctount 1804 Federal Documentary Microfilm Edition No. 1/3.

[73]See Robert Welsh Fenwick voucher 6/1/1827 Federal Documentary Microfilm Edition No. 1/4.

[74]See William Brown account 9/29/1827 5th auditor account 1957 Federal Documentary Microfilm Edition No. 1/4.

[75]See William Thornton to Henry Clay 3/6/1827 Federal Documentary Microfilm Edition No. 1/4.

[76]See Alexander M. McIntire voucher 9/2/1825 5th auditor account 1686 Federal Documentary Microfilm Edition No. 1/3.

Hotel after the 1814 burning of Washington.[77] Mr. Clay replied that while the Patent Office was attached to the State Department, it was inconvenient to have it located at a great distance from the rest of the department, as it would be if so moved. He indicated that he would support the creation of a Home Department to include the Patent Office, which would allow more flexibility in its location.[78] Blodgett's Hotel would soon be enlarged to secure the necessary room.

The Model Room became a popular tourist attraction. One German duke wrote of his tour in 1826, conducted by Jonas Keller, explaining in detail the kinds of models he found and his perception of the value of the various inventions.[79]

On March 28, 1828, Dr. William Thornton, Superintendent of the Patent Office since 1802, died at his home in Washington, D.C., following a long but relatively painless illness. In the first five years of issuing patents in the United States, an average of 20 patents per year were issued. In the five years immediately before Dr. Thornton entered the Patent Office, it averaged 45. In his first five years in office, Dr. Thornton averaged 73. In his last five years, the office over which he presided issued an average of 304 patents per year. When hired, his salary was $1,400 per year. After seventeen years, it had been raised to $1,500 per year. It was still $1,500 per year when he died. He was very near the end of his life before he stopped complaining about his salary.

[77] See John McLean to Henry Clay 3/5/1828 Federal Documentary Microfilm Edition No. 1/4.

[78] See Henry Clay to John McLean 3/7/1828 Federal Documentary Microfilm Edition No. 1/4.

[79] See [DoSWE28].

Chapter 13

Superintendent Thomas P. Jones, M.D.

The early history of Thomas P. Jones (1774–1848) is only vaguely known. He was probably born in Herefordshire, England, and was trained there as a physician. He was living in Philadelphia as early as 1796, having possibly emigrated with Joseph Priestly. By 1796, he was a member of a Unitarian congregation in Philadelphia founded by Priestly. Soon afterwards, he probably lived in New Bern, North Carolina. He delivered a series of scientific lectures in Albany, New York, and, by 1811, in Philadelphia. From 1814 to 1817, he was professor of natural philosophy and chemistry at the College of William and Mary. He resigned in 1817, with a public dinner given in his honor by his students, though possibly after an affray with a town person. Precisely who had the affray, or even if Professor Jones was involved, is not clear.[1]

He returned to Philadelphia in 1818 and resumed his course of technical lectures, including experimental lectures given at Charles Willson Peale's American Museum in the place of those previously given by C. W. Peale's son Reubens Peale. His assistant, slide-handler, and literally bottle washer during the museum lectures was C. W. Peale's grandson, who else, George Escol Sellers. He then left Philadelphia to run a school at Oxford, North Carolina. When the Franklin Institute was founded in February 1825, Dr. Jones was invited to return to Philadelphia to become its Professor of Mechanics and

[1]See *William and Mary Quarterly*, 26(1):140–41 (10/1917); *American Beacon and Commercial Diary* (Norfolk, VA) 9/13/1817 p 3 and 11/8/1817 p 3; William and Mary faculty minutes 12/9/1824.

Natural Philosophy and editor of its journal, which he accepted in March of that year.[2]

He returned in June and bought out the already existing American Mechanic's Magazine, which seems to have been a purchase of its subscription list and the right to continue its publication. This he published as the Franklin Journal and American Mechanic's Magazine, entirely at his own expense, or at least at no expense to the Franklin Institute. The effort of Peter A. Browne, mentioned earlier, to obtain access to patent specifications of unexpired patents was done to allow their publication in Dr. Jones' journal. In late 1827, Dr. Jones was offered a position on the faculty of the new University of Virginia. The Franklin Institute, concerned that he might accept, bought the rights to the journal from him, while giving him lifetime tenure as its editor. For whatever reason, he did not take the new faculty position.

Thomas P. Jones

Upon the death of William Thornton on March 28, 1828, there was a search for a new Superintendent of the Patent Office. According to a letter from Robert Peters of Philadelphia to Martin Van Buren,[3] Henry Clay had offered the position to a gentleman who refused it. That gentleman was Hezekiah Niles of Baltimore, who was offered the position on March 29.[4] Peters then recommended Dr. Jones to Clay, who appointed Jones to the position on April 12, 1828. This appointment came as a severe disappointment to William Elliot, who had expected that he would be appointed to the position,[5] and indeed he had a friend file a recommendation on his behalf on the day of Thornton's death.[6] Elliot wrote to apply for the position three days later.[7]

In announcing his removal to Washington, Dr. Jones announced that the journal would benefit from his new position. He said:

[2]See [Pre85], [Fow90], and papers of Pasquale Joseph and Bianca Morse Federico, Smithsonian National Museum of American History Library.

[3]See Robert Peters to Martin Van Buren 4/11/1829 National Archives M639 12:503–4.

[4]See Hezekiah Niles to Henry Clay 4/2/1828 National Archives M-179.

[5]See William Elliot to Martin Van Buren 3/10/1829 National Archives M-639 12:491.

[6]See Letter to Henry Clay from illegible sender 3/28/1828 National Archives M531 3:47.

[7]See William Elliot to Henry Clay 3/31/1828 Patent Office Letter Book mss p. 33 in papers of Pasquale Joseph and Bianca Morse Federico, Smithsonian National Museum of American History Library.

> The Repertory of Patent Inventions, The London Journal of Arts and Sciences, and The Register of Arts and Patent Inventions published in London, are principally devoted to the patents of Great Britain, whilst the more numerous patented inventions of this country have remained almost entirely unknown. When the Franklin Journal was first established, it was intended, among other objects, as may be seen by the prospectus in the first number, to embrace the patents of our own country, and several of them have been published accordingly; but the difficulty of making a proper selection, whilst the Editor resided in Philadelphia, was such as to preclude their frequent appearance; his present station is in this particular most eligible, and his determination is to turn this circumstance to good account. The list of American patents will hereafter appear regularly, so as to embrace, in each number, those obtained in the course of a month.[8]

After some difficulty in publishing the journal while moving, he succeeded for the first time in publishing a notice of all patents and a more detailed description of all important U.S. patents soon after they issued.

Dr. Jones came into the Patent Office intending to run the office his own way. He remained in the office after usual business hours, going over the recently issued patents, copying out interesting information from them, and using that information in his journal. He opened all incoming mail himself, without any assistance from the clerks in the office. Dr. Thornton had not done things this way. By October 1828, William Elliot had written a letter to the Secretary of State, complaining that Dr. Jones was impeding the work of the office by insisting upon opening all mail himself, with the result that, on days when Dr. Jones was absent, no mail was opened.[9] In March 1829, immediately after the inauguration of Andrew Jackson as President, with the promise of wide-spread removal of civil-service incumbents from office, William Elliot filed charges against Dr. Jones, beginning a full-scale frontal attack.[10]

Thomas P. Jones sent Secretary of State Martin Van Buren his response to Elliot's charges in May 1829.[11] There were many grounds for William Elliot's complaints against Thomas P. Jones, but the principal reasons, although unstated, were two. William Elliot did not like it because the Superintendent was not William Thornton, and, most important, William Elliot did not like it

[8]See [Jon28].

[9]See Daniel Brent to Thomas P. Jones 10/22/1828 National Archives M40 20:313.

[10]See William Elliot to James A. Hamilton 3/10/1829 National Archives M639 12:489.

[11]See Thomas P. Jones to Martin Van Buren 5/14/1829 National Archives M639 12:515–20.

because the Superintendent was not William Elliot. William Elliot had his son Seth Alfred Elliot employed at the Patent Office, although not by the Patent Office, making copies for the public at 20 cents per 100 words. William Elliot referred applicants who needed their applications written to his son John Elliot, who made out their specifications. The Elliot family had a good thing going at the Patent Office, sort of a family business. Dr. Jones came in and interfered with that business.

Jones wrote out his own copies of patents for his journal, despite Seth's offer to assist him. When unlettered and indigent persons came into the office seeking a patent, Dr. Jones would frequently write their specifications out for them on the spot, without charge, depriving William Elliot of opportunities to direct the applicants to his son John. Dr. Jones also said that he had written half a dozen specifications at his home for people who came to him because of their special trust in his ability. In response to Elliot's charge, he admitted that he received money for doing so. He claimed he tried to avoid it, not because he felt it was at all improper, but because he was too busy to devote time to writing patent applications. He said that at the request of applicants, he had drawings made for them by persons unconnected with the office at the lowest available price.

Elliot noted that during Dr. Thornton's tenure, many parties had sent presents of money to the Superintendent, and that all letters were opened in the clerk's room, where the gifts of money were removed and returned to the sender. He complained that Dr. Jones opened all letters sent to the Patent Office in private and that little or no money had been returned during Dr. Jones' tenure. Jones replied that he had yet to learn that it was his duty to open letters addressed to him in the presence of his clerks to keep him honest. He did not doubt that less money was sent as presents now than formerly, but he did not desire to discuss the probable reasons.

Elliot observed that Jones had issued patents without proper drawings, and in some instances without any drawings at all. Jones replied that he must be the judge of what were proper drawings, but that when he had consulted Elliot earlier, Elliot had advised accepting certain doubtful drawings, remarking that in some sections of the country, good drawings could not be obtained. Other charges, of less latter-day interest than the above, were also made and refuted.

Blodgett's Hotel after the 1829–1830 expansion

In the resolution of these charges, William Elliot came out second best, as he seems to have realized in advance. He wrote to Secretary Van Buren in late March,[12] requesting that if he should not be continued in office, one of his sons, William or Seth, be appointed in his place. Seth Alfred Elliot was a free-lance copyist at the Patent Office, and William Parker Elliot had been, for the previous two years, a student of architecture in London and Paris. By May, Mr. Van Buren had decided that William Elliot must vacate his position, but that William Parker Elliot would be allowed to be a free-lance draftsman in the Patent Office.[13] At about the same time, perhaps as late as November 1830, Robert Welsh Fenwick left his position as messenger of the Patent Office.[14]

Thomas P. Jones was transferred from his position in the Patent Office to another clerkship in the State Department, to be in charge of consular correspondence at the same salary. At about this time, he was also appointed Professor of Chemistry and Dean of the Medical Faculty at Columbian College (now the George Washington University). These, plus his continued position as editor of the Journal of the Franklin Institute, must have kept him occupied. The Franklin Institute supervised a massive letter-writing campaign to secure justice for Dr. Jones, but Dr. Jones nevertheless lost his job as Superintendent. He did continue to have virtually unlimited access to the records of issued patents in the Patent Office, allowing his continued publication of such patents in his journal.

But the Elliots had seen nothing to complain about until they saw the next Superintendent, Dr. John D. Craig.

[12]See William Elliot to Martin Van Buren 3/30/1829 National Archives M639 7:491–95.

[13]See William Parker Elliot to Martin Van Buren 5/11/1829 National Archives M639 7:505–6.

[14]See papers of Pasquale Joseph and Bianca Morse Federico, Smithsonian National Museum of American History Library

Chapter 14

Superintendent John D. Craig Inspires Protest

The next Superintendent of the Patent Office was Dr. John D. Craig (1766–1846), who was born in Ireland. There is a report from a frequently unreliable source[1] that although he was born in Ireland, his father was an American. The source and subject of his doctorate have not been discovered. In 1810, John and John D. Craig, perhaps father and son, were teachers at the Baltimore Union School.[2] From at least 1817 through 1824, John D. Craig was master of an academy in McClellan's Alley, Baltimore.[3]

In 1828, Dr. Craig gave a speech which led to the foundation of the Ohio Mechanics' Institute in Cincinnati,[4] which still exists in the OMI College of Applied Science of the University of Cincinnati. When he left the Patent Office in 1835, Dr. Craig returned to OMI as superintendent and librarian.

Thomas P. Jones was transferred to consular correspondence on June 10, 1829, and the following day John D. Craig became Superintendent. He immediately investigated the improprieties of his predecessors. He compared the number of patents issued in the past with the amount credited to the Treasury for the necessary fees. He found a shortage of over $4,000 occurring between 1815 and 1827. He requested explanations of the shortage from the widowed Mrs. William Thornton and from William Elliot and Robert Welsh Fenwick.

[1] See [Lan87].
[2] See Baltimore Directory 1810.
[3] See Baltimore Directory 1817–18 and 1824.
[4] See [Ken53].

All responded, but no satisfactory explanation was provided.[5] Some patents may have been issued to poor inventors without payment of fees. William Elliot said William Thornton did not believe it was the purpose of the patent system to amass revenue. Other patents may have been issued by mistake without payment of a fee. Mrs. Thornton brought her diaries into the State Department to show that they lived on a very small amount of money inconsistent with embezzlement. Dr. Craig said he did not intend to investigate the origin of the anomalies. He intended to prevent it from happening again by requiring a Treasury receipt to be attached to each proposed patent when reviewed by the Attorney General. The anomalies were never reconciled.

Early in the Jackson administration, while Dr. Jones was still Superintendent, Charles Bulfinch Jr. (1794–ca 1860), son of the departing Architect of the Capitol, began as a clerk in the Patent Office. By January 1, 1830, Dr. Craig was complaining to Secretary Van Buren that Bulfinch did only about 40 percent of the copying per day that a competent clerk should do.[6] Three years later, Dr. Craig was still complaining of the frequent absences and poor performance by Mr. Bulfinch.[7] By June 1833, Dr. Craig charged Bulfinch with dereliction of duty, claiming that poor health, real or imaginary, and bad eyesight left him unable to perform his ordinary duties. [8] Bulfinch replied,[9] denying that he was in poor health or that his eyesight was worse than others who wore glasses. He blamed his reduced production on his assignment to copy specifications onto parchment for issuing patents, which he considered more time-consuming than copying onto paper. By July, Bulfinch had been discharged, but a suspension of the order of discharge was allowed for a short time to allow Secretary McLane to review the order.[10] Bulfinch actually left office on November 30.[11] This left Dr. Craig free to appoint as his replacement, William Tell Steiger, a highly regarded former student, to be heard from later.

In apparently the first annual report on the condition of the Patent Office in December 1829,[12] Dr. Craig reported that new quarters for the Patent Office in Blodgett's Hotel were being constructed and furnished. He had by this time at least partly established the first regular system for arranging models and

[5]See [otCoP30]

[6]See John D. Craig to Martin Van Buren 1/1/1830 National Archives M639 3:694.

[7]See John D. Craig to Edward Livingston 1/1/1833 National Archives M639 3:696.

[8]See John D. Craig to Louis McLane 6/24/1833 National Archives M639 3:705–6.

[9]See Charles Bulfinch Jr. to Louis McLane 7/5/1833 National Archives M639 3:710–13.

[10]See Charles Bulfinch Jr. to Louis McLane 7/17/1833 National Archives M639 3:741–42.

[11]See Louis McLane to John D. Craig 11/11/1833 National Archives M40 24:91.

[12]See John D. Craig Report, *Condition of the Patent Office and Deficiency in its Accounts*, 12/22/1829 in House of Representatives Document 38 21st Congress 1st Session 1/27/1830.

drawings by subject matter. He was dissatisfied that issued patents had not been properly copied into record books as required by statute.

In his annual report on the condition of the Patent Office made January 1, 1831,[13] Dr. Craig wrote that the Patent Office quarters in the new section of the expanded Blodgett's Hotel had been furnished at a cost of $4,600. Craig wanted a machine shop for the Patent Office to be built on the second floor of the Engine House adjoining the Patent Office, but the act of Congress authorizing the building had authorized it exclusively for the engine, its apparatus, and the meetings of the Fire Engine Company. Dr. Craig considered the Engine House to be badly situated for use by the Fire Company and suggested a new building be erected in a convenient nearby location, allowing the old Engine House to be turned over to the Patent Office. He also indicated that only 900 of the 6,371 patents issued to date had been properly recorded by copying them into record books as required by law and suggested hiring ten clerks for two years to bring the recording up to date.

In an annual report made on December 2, 1831,[14] Dr. Craig indicated that all papers, drawings and models in the Patent Office were then classified and arranged so that what was sought could be found quickly. Apparently, this was the first time that the inventions available in the Patent Office were well classified by subject matter. In another annual report made January 1, 1834,[15] Dr. Craig indicated his belief that the framers of the patent laws did not intend them to be a source of revenue, but nevertheless, the Patent Office had generated and paid into the Treasury $110,000 in excess of its costs over the years. Dr. Craig wanted the money spent to build a suitable fireproof building for the Patent Office.

In the late winter and early spring of 1831, a number of clerks were appointed to copy the old patents into record books as required by the Act of 1793. One of the clerks so appointed was William Augustus Weaver (1797–1846), a former naval officer. He said that all of the clerks were housed in a barn of a room, without seats, entirely unheated in the depths of winter, at times with snow on the ground. Captain Weaver asked Craig that they be provided with fire and said he "was denied rudely and churlishly." He next asked Secretary Livingston, who then came to the Patent Office, indicated where stoves were to be put, and provided the necessary funds to Dr. Craig. Craig refused to provide the stoves, even after being ordered to do so by the

[13]See [otCoP29].
[14]See Document 2, 22nd Congress 1st Session.
[15]See [otCoP33].

Secretary of State and provided with the necessary money. He dismissed three copy clerks for not performing their work well under these conditions.[16]

An examination of the record shows that Dr. Craig accomplished a lot during his tenure, but he was arrogant, subject to rages, disagreeable to patent applicants and their agents, and a domineering tyrant toward the subordinate employees in the Patent Office. This was a man who, if he observed someone pouring oil on troubled waters, would set fire to the oil. His chief clerk was Alexander McIntire, who was considered the complete toady, and who was nicknamed "Iago" by his fellow employees.[17]

Dr. Thornton and Dr. Jones would read all incoming patent applications in order to locate applications pending before the Patent Office at the same time and which claimed the same invention, which were called interfering applications. All of these interfering applications were referred to arbitrators to determine which applications should be issued. Thornton and Jones had also attempted to mitigate the faults of the patent law by notifying an applicant of known prior art which would anticipate the invention disclosed in the application, then giving the applicant the choice of withdrawing his application and having his application fee returned if he wished to avoid having an unenforceable patent issued, or of insisting upon his patent in accordance with the law even though the Superintendent suggested otherwise. This help was not required by the Act of 1793, but was done to prevent uninformed applicants from wasting money and expectations on unenforceable patents. Dr. Craig would not even read the applications.[18] Copying clerks who were assigned to copy specifications onto parchment for use in the issued patent sometimes found the applicants' specifications illegible and uncopyable, showing clearly that they could not have been read. And indeed Dr. Craig said that he did not take the trouble to read specifications. He believed interferences could occur only between applications which were filed or put in condition for allowance on the same day, which virtually never happened. Five years into his tenure, Dr. Craig had never submitted any applications to arbitration for interference.[19] Craig believed that if an application was filed with all the proper papers and with the fees paid, a patent should issue, even if its contents were nonsense. The Act of 1793 intended that the courts should sort out such matters, not

[16]See Edward Livingston to John D. Craig 4/17/1832 National Archives M40 23:77; William A. Weaver to W.P. Elliot 12/28/1833 National Archives M639 7:524–26; Testimony of William A. Weaver 2/12/1834 Senate Document 398 23rd Congress 1st Session; and National Cyclopedia of American Biography 13:226.

[17]See Testimony of Henry Bishop 2/14/1834 Senate Document 398 23rd Congress 1st Session.

[18]See Testimony of Thomas P. Jones 2/15/1834 Senate Document 398 23 Congress 1st Session.

[19]See Decision of Dayton 3/15/1834 Senate Document 398 23rd Congress 1st Session.

the Patent Office, and Dr. Craig had no intention of doing more work at the Patent Office than the law required.

On October 23, 1833, about a month before the departure of Charles Bulfinch from the Patent Office, Dr. Craig wrote a letter to a former student, William Tell Steiger (1801–1888) of Baltimore, asking if he wanted to work at $800 per year as a clerk in the Patent Office.[20] Steiger was a former student of Dr. Craig, and was losing money in the leather and hide business in Baltimore. Dr. Craig had high regard for Steiger's ability. When Steiger wrote back to ask about the conditions of employment, Craig replied that office hours were from 9 to 3 o'clock, and that applicants frequently wanted drawings prepared for their applications, which Steiger could do out of office hours and might "realize something considerable" from doing so.[21] Craig recommended his appointment on November 8 to Secretary McLane,[22] calling him "a gentleman profoundly acquainted with mathematical and mechanical science, and in all respects well qualified to perform the duties of draftsman and clerk." Steiger began work on December 2.[23]

There are only scattered records of the history of the Patent Office during this period, but through a quirk of history the career of William Tell Steiger in the Patent Office can be seen exceedingly well. Steiger married Maria Shriver of the successful and politically active Shriver family of Maryland. He wrote frequent letters to his father-in-law, Andrew Shriver of Union Mills, Maryland, explaining what was going on in his career. The Shriver family preserved its vast correspondence and donated it to the Maryland Historical Society. Within a month of his appointment, Steiger was in competition with William P. Elliot for the private drafting business around the Patent Office.[24]

Since his father's departure from the Patent Office, William Parker Elliot had been allowed his own drafting table in the Patent Office, where he made copies of drawings for the Patent Office and made both original and copy drawings for private parties. He probably also wrote patent applications for private parties by this time. He was not an employee of the Patent Office but was paid by all parties, including the Patent Office, for work actually done. He claimed in June 1833 that he was never paid more than $600 per year by the Patent Office.[25] Dr. Craig wanted authority to have the office

[20]See John D. Craig to William Tell Steiger 10/23/1833 National Archives M639 7:566.

[21]See John D. Craig to William Tell Steiger 10/26/1833 National Archives M639 7:567.

[22]See John D. Craig to Louis McLane 11/8/1833 National Archives M639 23:64–65.

[23]See Testimony of William Tell Steiger 2/14/1834 Senate Document 398 23rd Congress 1st Session.

[24]See William Tell Steiger to Andrew Shriver 1/4/1834 manuscript papers of Shriver family, Maryland Historical Society, Baltimore, Maryland.

[25]See William Parker Elliot to Louis McLean 6/26/1833 Patent Office Letter Book mss p. 100.

drawing copies made only by an employed clerk, and he wrote several times to Secretaries Livingston and McLane for permission to give that work only to regularly employed clerks. Louis McLane directed in October 1833 that this should be done.[26] When Craig got rid of Bulfinch and had Steiger, a competent draftsman, in place beginning December 2, he immediately acted. By December 12, he had ordered Elliot to give up his drawing table, had been refused, and had written to Secretary McLane for directions. McLane informed him on December 16 that if the table was the property of the office, it was his to control. McLane reiterated that no clerk could be employed for his own private profit to prepare papers to be used by applicants or to copy papers on record in the Patent Office.[27] On the same day, McLane sent a set of charges of malfeasance and incompetence to Dr. Craig which Elliot had already filed against Craig.[28] And the battle was joined.

Steiger reported to his father-in-law[29] on January 4, 1834, that Elliot still held his table at the office and executed drawings for patent applicants. By March, Steiger, who was naturally a partisan for Craig, wrote a detailed letter to Andrew Shriver, giving him an explanation of the controversy in the office from a side not seen in the official record.[30]

According to Steiger, he was making drawings for applicants in his spare time, as Dr. Craig had promised him when he was hired. Secretary McLane had issued an order prohibiting it, but Dr. Craig told Steiger there was nothing in the law to prohibit it. Elliot wrote a letter to the Secretary, complaining that Steiger was doing drawings outside the office (and in competition with him). After Elliot filed charges against Craig, Craig demanded a hearing. When the hearing was not offered promptly, he sued Elliot for slander and libel. When President Jackson heard of the suit, he told the Secretary to order the removal of Elliot from the office and to make arrangements for the hearing. Steiger wrote that the questions put in the hearing were entirely irrelevant to the charges. At first he refused to answer questions concerning his out-of-office work, but he answered after receiving express orders to do so from Secretary McLane. He wrote that if he were fired, he could earn more money in the time now spent working for the Patent Office by doing private drawings for the applications of inventors. He thought that Elliot was an insincere and false man who was a tool of someone else, perhaps Dr. Jones. He thought that Dr.

[26]See Louis McLane to John D. Craig 10/17/1833 National Archives M40 24:77.

[27]See Louis McLane to John D. Craig 12/16/1833 National Archives M40 24:117.

[28]See Louis McLane to John D. Craig 12/16/1833 National Archives M40 24:116.

[29]See William Tell Steiger to Andrew Shriver 1/4/1834 manuscript papers of Shriver family, Maryland Historical Society, Baltimore, Maryland.

[30]See William Tell Steiger to Andrew Shriver 3/5/1834 manuscript papers of Shriver family, Maryland Historical Society, Baltimore, Maryland.

Craig was a favorite of President Jackson, since Jackson had ordered that the evidence should come before him and that justice be done to Dr. Craig.

According to the official record,[31] depositions of everyone associated with the Patent Office and several outsiders were taken from February 12 through 24, 1834 by Aaron Ogden Dayton (1796–1858), the officer appointed by the Secretary to take the evidence. Findings were presented on March 15 by Mr. Dayton to the Secretary. Craig asked that Mr. McIntire be allowed to represent him in the hearings and that he not be required to attend. Charles M. Keller, the son of Jonas Keller who had been an unpaid youthful repairer of models in the 1822 Patent Office, was by now the machinist of the Patent Office after the death of Jonas. Machinist Charles Keller testified that Senator Gabriel Moore had complained to him of being treated rudely by Superintendent Craig. He also stated that there were pieces of models which could not be identified, for which no drawings could be found, which may not have ever been patented, and which could not be repaired. These pieces had been around since his father was machinist. Dr. Craig ordered the destruction of the pieces, and Keller willingly destroyed them until ordered to stop, since he could think of absolutely no reason for keeping the pieces.

Another witness was Henry Bishop, successor to Robert Welsh Fenwick as messenger of the Patent Office. He testified that he had heard visitors complain of rudeness toward them by Dr. Craig, and that he had also heard Dr. Craig complain of rudeness toward him by visitors. He thought that Dr. Craig was very diligent in carrying out his duties. Bishop said that Elliot had once complained that, although Dr. Jones had been an attentive Superintendent, he had taken fees that were improper. He also said that since the falling out between Craig and Elliot, Elliot and Jones seem to have become much better friends than formerly.

Dr. Jones testified that Craig would not declare arbitrations and indeed would not even read the specifications of patents he was about to issue. He also said that Craig had told him that the order of the Secretary of State forbidding clerks from executing drawings for individuals for their private profit was a tyrannical order. Steiger and Weaver testified to matters as mentioned above, then Alexander McIntire testified in general defense of Dr. Craig.

Mr. Dayton found that Dr. Craig had indeed failed to declare interferences where it should have been done, that he had discontinued keeping the Caveat Book without which it would be impossible to carry out his duties properly, that he had unnecessarily and imprudently allowed correspondence of the office to be destroyed which had been preserved by previous Superintendents, that he had clearly been rude toward visitors (as had even been observed personally

[31]See Senate Document 398 23rd Congress 1st Session 5/26/1834.

by Dayton in the course of his investigation), that he had given applicants incorrect information from a misunderstanding of the law, that he had been inattentive to duty in not reading specifications before issuing patents, that he was charging a higher price than allowed by law for copies of drawings, and that he allowed Steiger to make drawings out-of-office for private profit in violation of the Secretary's order, which he called tyrannical.

Secretary McLane wrote a letter to John D. Craig on March 28, 1834,[32] to tell him that there was much in his conduct which was deserving of censure, but that he would refrain at the present from doing more than to order him to correct certain enumerated deficiencies or be subject to removal. William P. Elliot wanted to know the result of his charges, and he asked for a copy of the report on April 1, which was refused a few days later, although Elliot was given a copy of the March 28 letter to Craig.[33] Elliot then wrote directly to President Jackson, who instructed McLane to refuse.[34] Perhaps the President thought that would end the matter, but it did not. Apparently, Elliot reached someone in the Senate and persuaded the Senate to request that copies of the entire proceedings be forwarded there. This should not have been difficult, knowing that among those people whom Dr. Craig honored with his rudeness was Senator Gabriel Moore. The copies were forwarded and printed, but Jackson would not forgive William P. Elliot this insult to his authority.

Apparently, Alexander McIntire filed for relief under the Insolvency Laws[35] and was notified in July 1834 that he was to be dismissed from the Patent Office.[36] The real reason may have been different. He was removed in August, obviously contrary to the desires of Craig.

His replacement, appointed contrary to the desires of Dr. Craig, was Dr. Robert Mayo (1784—1864), a physician and author from Richmond, Virginia. He was a Jackson supporter, having edited the Jackson Democrat in Richmond during the 1828 campaign. He later came to Washington as a civil servant. Upon learning of his appointment, Dr. Craig wrote him a welcome-to-the-office letter, stating that Mayo had been said to have a disagreeable temper, to try constantly to undermine and supplant others, and to be a very inferior clerk.[37] It seems to have been generally believed that Mayo was a spy planted by Jackson in the Patent Office to find out what was happening.

[32]See Senate Document 398.

[33]See Louis McLane to William P. Elliot 4/5/1834 National Archives M40 24:217.

[34]See William P. Elliot to Andrew Jackson 4/9/1834 National Archives M639 7:534–36; LM to WPE 4/11/1834 National Archives M40 24:224.

[35]See papers of Pasquale Joseph and Bianca Morse Federico, Smithsonian National Museum of American History Library.

[36]See John Forsyth to Alexander M. McIntire 7/1/1834 National Archives M40 24:286.

[37]See John D. Craig to Robert Mayo 7/3/1834 National Archives M639 15:975.

The quarrel continued. William P. Elliot went to the Patent Office to inspect some recently issued patents. When he asked Dr. Craig for access, Dr. Craig remained silent. He then sought access from Steiger, who told him that he could not speak to him. Elliot wrote to Secretary John Forsyth to complain.[38] Elliot was soon filing new charges against the Patent Office that Dr. Craig was allowing William Tell Steiger to make drawings for private payment. Mr. Steiger in fact made $600 for private drawing in 1834.[39] By January 1835, Dr. Craig was filing charges against Mayo, accusing him of being a spy for William P. Elliot and of having filed for relief under the Insolvency Laws.[40]

If you are getting tired of this bickering, you can see why the Secretary of State, undoubtedly with the approval of President Jackson, notified John D. Craig that he was dismissed. This was said to be because of his expressed belief that the rule forbidding clerks to receive money for assisting applicants was tyrannical.[41] Steiger regretted that the poor old man had been dismissed and thought it highly probably that he would soon be dismissed himself. His one consolation was that Dr. Mayo had been dismissed at the same time.[42]

[38]See William P. Elliot to John Forsyth 11/18/1834 National Archives M639 7:547–48.

[39]See William P. Elliot to John Forsyth 11/24/1834 National Archives M639 7:554–56; William Tell Steiger to John Forsyth 11/29/1834 National Archives M639 7:560–63; Thomas P. Jones to John Forsyth 12/1/1834 National Archives M639 7: 570–73 William Tell Steiger to Andrew Shriver 12/28/1834 manuscript papers of Shriver family, Maryland Historical Society, Baltimore, Maryland.

[40]See John D. Craig to John Forsyth 1/7/1835 National Archives M639 15:982–84; RM to JF 1/10/1835 National Archives M639 15:985–88.

[41]See John Forsyth to John D. Craig 1/31/1835 National Archives M40 25:207.

[42]See William Tell Steiger to Andrew Shriver 2/1/1835 manuscript papers of Shriver family, Maryland Historical Society, Baltimore, Maryland.

Chapter 15

The Old Order Prepares to Change

Col. James Chamberlayne Pickett (1795–1872), a genuine Kentucky colonel and former diplomat in Columbia, was appointed Superintendent of the Patent Office on February 1, 1835, to replace John D. Craig.[1] He was in office only a short time, being appointed fourth auditor of the Treasury on May 1, 1835,[2] and he later served seven more years as a diplomat in Ecuador and Peru. In his exactly three months in office, he set a new record. No charges of any kind were filed against him by anyone, the first Superintendent who could make that statement. Very shortly after Pickett came into office, Robert Mills (1781–1855), the architect, was appointed a clerk in the Patent Office to the vacancy left by the departing Dr. Mayo.[3]

The Administration had a hard time deciding what to do with William Tell Steiger. He was an excellent draftsman, which was a skill in short supply. He had defied the Secretary by continuing to do private drawings, but he had been hired with a promise that he could do so, and Dr. Craig had told him to the end that he could do so. His removal from his $800-per-year position as Patent Office draftsman was ordered on April 6, effective the next

[1]See National Cyclopedia of American Biography 13:159; William Tell Steiger to Andrew Shriver 2/1/1835 manuscript papers of Shriver family, Maryland Historical Society, Baltimore, Maryland.

[2]See John Forsyth to James Chamberlayne Pickett 5/1/1835 National Archives M639 18:425.

[3]See John Forsyth to James Chamberlayne Pickett 2/7/1835 National Archives M40 23:214.

day.[4] Colonel Pickett immediately promised to search for a new job for him if he would not give vent to his feelings on the subject. Dr. Jones immediately gave him $50 worth of patent drawing work that he had been afraid to give him earlier while he was still employed in the Patent Office and offered to circulate his advertisement in the Journal of the Franklin Institute in order to get him more work.[5] In three weeks, Steiger had earned $160 as a totally private draftsman, getting work referred to him by Jones, Pickett and members of Congress. Meanwhile, Robert Mills had been delegated to be the Patent Office in-house draftsman. Steiger wrote that Pickett was having difficulty getting his work done by Mills, and Steiger's former duties were divided between Mills, Pickett, Keller, and a newly appointed clerk, Thomas Johns. By July, Pickett and the President had obtained the job of draftsman of the General Land Office for him at $1,150 per year, with no restrictions on his ability to do patent drawings on private time. Steiger lived across the street from the Patent Office and had no difficulty continuing his patent work. Charles Michael Keller (ca 1809–1874), who had been working at the Patent Office since 1822 had by 1835 the longest service in the Patent Office by far. Even though he was only about 25, he had spent over half his life in the Patent Office. When his father died in 1830, he had been appointed machinist in his father's place, even though he was still a minor. Keller proposed to Pickett some revisions in the law and the practice of the office to eliminate the practice of deliberately or accidentally patenting the same invention more than once. Pickett assigned Keller the duty of advising applicants concerning the novelty of their inventions.[6] Colonel Pickett left too quickly for any other action on Keller's suggestions.

James C Pickett

Henry Leavitt Ellsworth (1791–1858) was one of the twin sons of Chief Justice Oliver Ellsworth. His twin William Wolcott Ellsworth was Governor of Connecticut. Henry had been chief commissioner to the Indian tribes to

[4]See John Forsyth to James Chamberlayne Pickett 4/6/1835 National Archives M40 25:300.

[5]See William Tell Steiger to Andrew Shriver 4/6/1835 manuscript papers of Shriver family, Maryland Historical Society, Baltimore, Maryland.

[6]See [Edi59b].

the south and west of Arkansas and made several journeys towards the Rocky Mountains. On one of his trips, he was accompanied by Washington Irving, who wrote of the journey in his Tour of the Prairies. From Ellworth's exposure to the West and knowledge of inventions derived from time in the Patent Office, he prophesied late in life that the lands of the West would be cultivated by means of steam plows. This prophecy was introduced in the probate of his will in an attempt to prove that he was of unsound mind.[7]

Henry Ellsworth applied in January 1835 to be Commissioner under the Spanish treaty.[8] Apparently this application was not entirely fruitless, because he was appointed Superintendent of the Patent Office and took the oath of office on May 11. He immediately requested and received a 60-day leave of absence to arrange personal matters before reporting for duty.[9] Thomas Johns was acting Superintendent until Ellsworth reported.[10] By July 4, Steiger was speculating that he might be made Superintendent if Henry Ellsworth did not arrive.[11] But Ellsworth did report on July 8. Steiger said that both Thomas Johns and Robert Mayo had already applied for the position in the expectation that Ellsworth would not arrive.[12] He thought they had been lighting on the carcass before the life was out and found this similar to the case of Samuel C. Potter, an agent for the Treasury Department, who had been one of 400 persons on a three-day excursion by boat on the river over the 4th of July holiday. He was apparently drunk and rolled overboard at 2 or 3 o'clock on Monday morning and drowned. The boat did not return to Washington until 9:30 Monday morning, but applications for his position had started to arrive at the Treasury Department by nine!

A month after his arrival, Ellsworth was explaining what he had been doing with his time. He said that all of the recent correspondence which had not been destroyed by Dr. Craig was kept strung up on two long wires, and every time it was necessary to consult papers relating to an application, it was necessary to search through the unindexed papers. Each time a letter from the wires was consulted and put back on the wires, a new hole was pierced, obliterating part of the copy. Ellsworth was having all of the letters bound and indexed to make it quicker to consult them. He found a third of the floor in the Superintendent's

[7]See National Cyclopedia of American Biography 7:516.

[8]See Henry Leavitt Ellsworth to John Forsyth 1/26/1835 National Archives M639 7:610–11.

[9]See Henry Leavitt Ellsworth to John Forsyth (3 letters) 5/11/1835 and 5/18/1835 Patent Office Letter Book p. 245–46.

[10]See Johns to Dickens 5/29/1835 Patent Office Letter Book p. 246.

[11]See William Tell Steiger to Andrew Shriver 7/4/1835 manuscript papers of Shriver family, Maryland Historical Society, Baltimore, Maryland.

[12]See William Tell Steiger to Andrew Shriver 7/9/1835 manuscript papers of Shriver family, Maryland Historical Society, Baltimore, Maryland.

office occupied by over 60 models of unpatented inventions and removed them to a separate room, making the Superintendent's office much easier to use. He found that no list of applicants for patent had ever been made, which deficiency he corrected. He also had envelopes printed, one of which was to be used for each application, creating the first application files. Although he considered himself almost caught up on the work as far as strangers were concerned, he was far behind on the internal reports that were considered necessary. He thought that some revision to the patent law should be made to eliminate the need for signatures for so many officials, including the President. He noted that 50 patents were then awaiting the signature of the President.[13]

Henry Ellsworth

The next month, Ellsworth explained that the very building was inadequate. There was immediate need for more space to store models, and several hundred were piled away in the garret of the Post Office. The great weight of the models was endangering the building. Ellsworth noted that there was a surplus of $130,000 to the credit of the Patent Office in the Treasury, in what was later called the Patent Fund. He suggested that this money would be adequate to build a suitable building for the Patent Office.[14]

John Gill (1798–1843), of New Bern, North Carolina, a locksmith, watch-maker, goldsmith and silversmith, made his first revolving gun in 1829. It was a predecessor of the six-shooter, but might more properly be called a fourteen-shooter, since it had 14 chambers. He was a man with very little money whose friends advised him not to waste it on a patent. He traveled by boat from New Bern up to Norfolk, then up the bay to Baltimore. Upon arrival in Baltimore he became ill before he could go down to Washington and the Patent Office. A visitor called on him while he was recovering and borrowed his model to show a friend. When he recovered and traveled to Washington, Mr. Gill found that Samuel Colt had arrived before him to patent his six-shooter. He was convinced that somehow Samuel Colt had learned of his invention. Before his death at an early age, Mr. Gill gave his tools and

[13]See Henry Leavitt Ellsworth to John Forsyth 8/10/1835 Patent Office Ltr Bk mss p 230–34.

[14]See Henry Leavitt Ellsworth to John Forsyth 9/29/1835 Patent Office Ltr Bk mss p 236–39.

mechanical books to a friend to hold in trust for Gill's two sons. Mr. Gill's original model was still owned by the Matthews family of New Bern in January 1861, but it was stolen by Union solders during the 1862 capture of New Bern. Remarkably, after the Civil War, John Gill's son Richard C. Gill obtained a position at the Patent Office and by the 1880s was in charge of the model rooms, at that time the most successful museum in Washington.[15]

Model Room of the Patent Office about 1835

Charles M. Keller enlarged and refined the suggestions for improving the Patent Office which he had earlier given to Pickett and then presented them to Superintendent Ellsworth. A newly appointed Senator arrived from Maine, in the days when senators were still appointed by the state legislatures. Senator John Ruggles, who was an inventor himself, came to the Patent Office shortly after his arrival in Washington. The reason for his visit is not now remembered, but it was probably in connection with his invention for a cog railway. He met Mr. Keller, and they had a long discussion on the faults in the current patent system.

[15]See [Edi61c] and *New Bern Sun-Journal* 7/2/1976 p. 11E.

Chapter 16

1836 at the Patent Office—The Best of Years

The year 1836 was a watershed in the history of the Patent Office. Over a century and a half later, it is hard to realize what Washington City was like so long ago. The railroad was just coming in from Baltimore, and the 40-mile trip could now be made in comfort in only three hours. But the Steigers found it easier to get to Union Mills by canal boat because the boats stopped closer to Union Mills than did the trains, and it was hard to make their own way the rest of the distance when they had their small children with them.[1]

There were a dozen men landscaping the grounds of the Capitol. The Postmaster General was digging out the area around the front of Blodgett's Hotel, installing windows and an iron railing, so that the basement could be used for rooms for clerks.[2] Congressmen were paying $10 a week to board in local boarding houses, and during sessions of Congress, local people who boarded also paid these high rates.[3]

The Steigers had been heating their home with coal burned on open grates in a fireplace, but in November 1835 they bought one of Spoor's fine patent anthracite stoves, and when they burned coal in it in a first experiment, they

[1]See William Tell Steiger to Elizabeth Shriver 8/1/1835 and Maria Steiger to Elizabeth Shriver 9/25/1835, manuscript papers of Shriver family, Maryland Historical Society, Baltimore, Maryland.

[2]See Maria Steiger to Elizabeth Shriver 2/24/1834 and William Tell Steiger to Elizabeth Shriver 8/11/1836, manuscript papers of Shriver family, Maryland Historical Society, Baltimore, Maryland.

[3]See Maria Steiger to Elizabeth Shriver 12/4/1836, manuscript papers of Shriver family, Maryland Historical Society, Baltimore, Maryland.

had to open the windows to reduce the heat. They could now be comfortable in their home in winter, thanks to a patented stove.[4] On New Year's Day 1836, the Steigers went to President Jackson's levee and shook hands with the President, then rode a carriage home.[5] Many people in the city kept a cow to furnish their families with milk, and the city kept a bull in each ward of the city to serve the cows of the residents. Each householder was required to keep a number of buckets on hand for use of bucket brigades to put out fires. Ladders and pumping engines were kept in each ward for the use of the inhabitants.

Although hogs were allowed by law to run free in the northern part of the District, any hog allowed to go at large south of Massachusetts Avenue was to be seized and sold at auction, half of the proceeds going to the person doing the seizing and half to the city. The owner of any horse, cow, etc., which died on the streets was required to bury it within twelve hours or pay a penalty. Anyone cleaning fish at the city pumps was to be fined, as was anyone racing a horse on the city streets. Schools were maintained by the city.[6]

John Ruggles

The streets of the city were largely of dirt and mud. The city would not improve the streets because the citizens felt that the federal government should pay at least some of the expense. Congress would not appropriate money to improve streets for the people of the federal district. Hogs roaming the streets were in hog heaven. Garbage littered the streets, and mud made them feel at home. The street in front of Blodgett's Hotel was a favorite hog slopping ground. On April 14, 1836, a citizen was descending the steps from the Post Office in the middle of Blodgett's Hotel when a full-grown hog came running past, ran against him, and threw him to the ground, striking his skull against the curbstone. His skull was fractured, and he died two days later.[7]

[4]See William Tell Steiger to Elizabeth Shriver 11/6/1835, manuscript papers of Shriver family, Maryland Historical Society, Baltimore, Maryland.

[5]See Elizabeth Shriver to Andrew Shriver 1/1/1836, manuscript papers of Shriver family, Maryland Historical Society, Baltimore, Maryland.

[6]See [Ell22].

[7]See [Pro39].

On March 20, 1833, the Treasury Building was destroyed by fire. This greatly increased Congressional interest in housing the various government departments in fireproof buildings.[8]

John Ruggles (1789–1874) was born in Westover, Massachusetts, graduated from Brown University, and read law with Levi Lincoln in Worcester, Massachusetts. In 1817, he moved to Thomaston, Maine to practice law. He was soon elected to the Maine legislature, then was Speaker of the Maine House of Representatives, and was later a Justice on the Maine Supreme Court. In 1835, he was appointed by the Maine legislature to be a U.S. Senator from Maine.

On December 31, 1835, Senator Ruggles made a speech to the Senate to propose that a Senate Patent Committee be set up.[9] He said that he had recently had business to transact at the Patent Office, and while there had inquired into the causes of delays in issuing patents. His inquiry had revealed to him a number of defects in the patent laws, which had not been amended for nearly half a century except for extension of rights to certain foreigners. Patents were issuing at the rate of 800 a year, with an increase to 1,000 a year expected within a year. Back when 50 patents a year were issued, it may have been appropriate to have each patent signed by the Secretary of State, the Attorney General, and the President of the United States, but by 1835 it took three months to get all of the necessary signatures to issue a patent. Because the Patent Office officials had no discretion to refuse a patent for an invention which was not new or useful, applicants had been known to copy inventions on display in the model room and to request and receive a patent, despite knowledge of both the applicant and the Superintendent that the invention was not new. When the resulting patent was divided into license rights by states and counties, and the licenses for the fraudulent invention parceled out to ignorant buyers, who were shown documents signed by the President with the Great Seal of the United States, significant money could be made. Such fraudulent sales of licenses were estimated to amount to half a million dollars a year. The proposed committee was set up, and Mr. Ruggles was made its chairman.

Mr. Ruggles made a formal inquiry to the Secretary of State for information on how the Patent Office should be reorganized. On January 29, 1836, Mr. Ellsworth wrote a long responsive letter to Secretary Forsyth, setting forth all of the defects and suggesting corrections. The original letter was forwarded to Mr. Ruggles, who evidently regarded it as one of the important souvenirs of his term in the Senate. He kept the original letter, and it was among the few

[8] See [Bry16].

[9] See [Kni91a].

patent documents among his possessions when they were finally auctioned in 1992.[10]

In his letter, Mr. Ellsworth wrote that initially the poverty and distress of the country required that any law for the encouragement of inventions must be self-supporting. In other words, the full expenses of operating the patent system had to be paid for by the inventors themselves, since the country had no other money to appropriate for the patent system. This worked so well that, in spite of the great delays in obtaining patents because of lack of necessary personnel and space, the fee of $30 per application was large enough that the whole expenses of the Patent Office did not exceed one quarter of its income. But the Patent Office still had financial problems because all of the money received had to be paid into the Treasury, and it could not be spent without appropriation by Congress, which seemed to treat the patent fees as if they were general income of the federal government.

The Act of 1793 had intended the Patent Office to issue any patent properly applied for, and to leave it to the courts to sort out who should have what rights. But matters had reached such a stage that, while 800 patents a year were being granted, there were currently more than 100 suits pending in court regarding the rights of patentees. Mr. Ellsworth predicted that such suits could only increase until some method were provided to stop the frauds now openly practiced on the patent system.

Delays in the office concerned many members of the public. It was a common practice for an inventor to travel to Washington from a great distance, expecting to apply for a patent and carry the issued patent home with him. Others urgently needed copies of papers or reissued patents for use in court. Because patents had to be signed by the President, the Secretary of State, and the Attorney General, any of whom could be busy with matters more urgent and pressing than the issuance of a patent, there could be very long delays for the necessary signatures for issuing a patent. Many patents were transmitted 200 miles to get a single one of the necessary signatures. If the signatures of only the head of the Patent Office and the Secretary of State were required, the average time to issue a patent might easily be reduced from two or three months to a few days.

The clerks in the Patent Office were considered to be underpaid, and were mostly temporary clerks who did not have the mechanical and legal skills necessary for the job. The Superintendent himself was by law merely a clerk, and even at that was paid less than the Chief Clerks of the Auditor's Office, the Land Office or the Indian Bureau. The salaries of the lower clerks were 30

[10]See Henry Leavitt Ellsworth to Forsyth 1/29/1836, printed in *Mechanic's Magazine* 8:175–82 (1836).

to 50 percent less than in those other offices. One third of the revenue derived from patent fees would suffice to make the salaries equal, leaving two-thirds to be appropriated as Congress might direct.

The issuance of patents to pirates was a common occurrence, and indeed the Patent Office furnished every facility to aid them. Copies of models of previous inventions were made in the model rooms and used to demand patents for similar inventions next door in the Superintendent's Office. Even when an applicant was told that his invention was a direct copy of a previous invention, as had happened the previous week, he demanded and obtained his patent for the same invention. With the Great Seal and the signature of the President on his pirated patent, the pirate patentee could sell state, town and county rights all over the country to people who thought that the Great Seal and the signature of the President actually meant that the patent was valid. The present system provided a rich harvest for lawyers, but ruin for many an honest mechanic.

Jefferson had suggested in 1813 that the head of the Patent Bureau be given discretion to prevent the issuance of a patent which interfered with an earlier patent or which was destitute of novelty. Mr. Ellsworth repeated the suggestion. He wanted scientific men to be induced to take positions in the Patent Office as Examiners of Patents, to have a suitable library to aid them in their examination of applications to detect interference with other patents or want of novelty. Mr. Ellsworth had recently found in a German book a complete anticipation of a recently issued U.S. patent. He wanted the Patent Office to have a good library to aid in examination of patents.

In a comparison of the merits of the United States patent system with that of England, Mr. Ellsworth pointed out that in the last 60 days he had issued more than 200 patents, which was greater than the annual average in England for the last 10 years. England allowed patents for newly imported inventions, but Mr. Ellsworth considered that the reading spirit of the people of the United States was such that it was evidently better to confine patents here to new discoveries. The temptation to patent in this country was such that it might be well to compel each patentee to publish, if not his entire specification, at least his specific claim of novelty. We could scarcely eat, drink, sleep or work without using some patent. Does this remind the reader of a passage, from an earlier chapter, from Patents all the Rage—"Our clothes, our physic, and our food, with many queer utensils, must all be marked with patent stamps like warming pans and pencils."?

Mr. Ellsworth recommended that a small portion of the surplus money obtained from patent applications be used to publish all specifications of patents, or at least the claims under the patents. If the material thus published were distributed to the different states, the public would be well capable of guarding against spurious patents.

He wrote that the present Patent Office building was shared with the Post Office Department and was too small for their joint use. The Post Office needed the entire building, while the Patent Office had hundreds of models stored away in the garret for lack of sufficient room to display them. While more room was desperately needed, Ellsworth was particularly anxious that the same should be as secure as possible from fire, since the destruction of the present models would produce a very great embarrassment, especially since so many original patents and assignments had already been lost over the years. There was currently $150,000 to the credit of the Patent Office in the Treasury which should be available to build a permanent Patent Office Building. Mr. Ruggles had asked what additional expense would be incurred in providing an exhibition of machinery and works of art not patented. Mr. Ellsworth replied that with a suitable building having sufficient room, there should be very little additional expense.

Patents were not then allowed to non-resident foreigners. Mr. Ellsworth recommended that foreigners should be allowed to take out U.S. patents, but for a fee corresponding to that which their countries charged U.S. citizens for taking out patents in their countries. This was about $550 to 600 in England and Ireland, down to about $200 in France and Austria, compared with $30 in the United States.

This letter was apparently derived from the suggestions that Keller had made to Ellsworth. Working with this letter and their own ideas, Senator John Ruggles and Patent Office machinist Charles M. Keller worked together late into the night for many weeks to write a new patent law.[11] Mr. Ruggles, former Justice of the Maine Supreme Court, taught Mr. Keller a new respect for jurisprudence which led later to a career in patent law. Mr. Keller, experienced for years since childhood in dealing with inventors visiting the Patent Office, must have taught Mr. Ruggles some of the intricacies of dealing with inventors and inventions. For then as now, although most inventions are mere extensions of what was known earlier, it is the nature of invention that you must expect the unexpected. By late April 1836, the select committee had prepared its report on the Patent Office.[12] After reviewing the history of foreign patent laws and of the current U.S. patent statute, they discussed the frauds being committed because of the inability of the Patent Office to refuse applications without merit. They recommended examination of patent applications and the raising of the new organization above a mere clerkship. The previous system may

[11] See [Edi59b].

[12] See Senate Document 338 24th Congress 1st Session 4/28/1836.

have been good enough for the agricultural country that founded it, but it was not sufficient for the manufacturing nation which had arisen through American ingenuity and intellect. Who could predict what the spirit of enterprise and genius of invention had in store for the country? But agriculture might as well dispense with the fertility of the soil as with the newly invented aids for its cultivation. Mr. Ruggles reported:

> Whoever imagines that, because so many inventions and so many improvements in machinery have been made, there remains little else to be discovered, has but a feeble conception of the infinitude and vastness of mechanical powers, or of the unlimited reach of science. Much has been discovered, infinitely more remains unrevealed. The ingenuity of man is exploring a reign without limits, and delving in a mine whose treasures are exhaustless. Neither are all the mysteries of nature unfolded, nor the mind tired in the pursuit of them.

He recommended that foreigners be allowed to get patents here by paying fees comparable to what Americans would have to pay in their countries. He recommended that a Commissioner of Patents be empowered to reject applications for want of novelty in the invention, with appeal to arbitrators appointed for the occasion and called a board of examiners.

He recommended larger rooms for models to prevent them from being consigned, as 500 already had been, to the dark garrets of the Patent Office, to be unavailable for exhibition. He recommended spending a portion of the $152,000 surplus funds of the Patent Office to construct a building suitable for such display and comparatively safe from fire. The Patent Office, he said, contained the records of this age of inventions, which "would present an object of interest, and tend not a little to elevate our national character. It has been justly remarked that we can go into no mechanic shop, into no manufactory of any description, upon no farm or plantation, or travel a mile on our railroads or in our steamboats, without seeing evidence of our originality and witnessing the fruits of our ingenuity and enterprise." No other nation had ever required models with patent applications, and there was no comparable material for display available anywhere else in the world. A reorganization of the Patent Office was recommended, with a suitable museum of the arts to display models now boxed up in a garret for lack of room to display them. With these objects in mind, the select committee presented a draft statute to establish a Patent Office under the Department of State, to be headed by a Commissioner of Patents, to have an extensive library of scientific works and periodical publications, and, in addition to the necessary clerks, draftsman, machinist and messenger, to have an examiner of patents, the first anywhere in the world, to read patent applications, compare them with what was known in this country or contained in a printed publication anywhere prior to the applicant's discovery, or in public use or on sale with the applicant's consent prior to the application, and to reject those applications in which the invention was not new. The fee for an application was to be $30 for citizens and resident aliens, $500 for subjects of the King of Great Britain, and $300 for all other aliens. Models were still required, and provision was made for their public display.

A statute as requested by the select committee was passed by Congress and signed into law on July 4, 1836.[13] Again, the builders builded better than they could have known at the time. This statute, sixty years to the day after the original Declaration of American Independence, was a second such declaration. Some historians have said it was the most important thing to happen in the country between the War of 1812 and the Civil War. The first Declaration of Independence led to our political independence from England. The second led, just as inexorably, to our industrial independence from Europe. It gave this country such an overwhelming advantage toward national industrialization that European nations did not at first understand what had happened. American industrial progress was attributed to the native mechanical abilities of Americans. But as Abraham Lincoln was later to say, "the patent system added the fuel of interest to the fire of genius." Now, much more than ever before, it was possible for an inventive genius to profit by making and promoting an invention, which of course induced men to do so. By the beginning of the twentieth century, most of the industrialized countries in the world had copied significant parts of a statute which was written by a lawyer with an interest in his own invention and by a boy genius who started working in the Patent Office because his father could not afford to give him a good formal education and had to teach him by example.

On July 5, 1836, Mr. Ruggles wrote to Secretary John Forsyth, adding his recommendation to those of many others that Charles M. Keller should be appointed to the office of Examiner for the Patent Office.[14] Mr. Keller was promptly appointed. None of the patents which had issued prior to the act of 1836 had been given patent numbers, being referred to only by their date of issue. It was determined to number patents as they were issued under the new patent law. To honor Senator Ruggles, the first patent issued under the Act of 1836 was U.S. Patent number 1, issued to John Ruggles for a locomotive steam engine for use on inclined planes. This was probably the cog railway which

[13]See 24th Congress 1st Session 1836, Statute I, Chap 357.

[14]See John Ruggles to John Forsyth 7/5/1836 National Archives M687 18:67.

prompted his visit to the Patent Office in 1835. Years later, his granddaughter gave the original patent to Brown University.

Mr. Ruggles also proposed a bill on June 15, 1836, to construct a building for the accommodation of the Patent Office. After efforts failed to amend the bill to require the purchase of the Old Brick Capitol from the private citizens who owned it for use by the Patent Office, the Senate passed the bill to appropriate $108,000 to construct a building out of brick and wood. The House amended the bill to require the construction to be of cut stone and fireproof inside and out. The bill was passed by Congress and was signed into law on July 4, 1836.[15] Both the House and the Senate approved the plan submitted by William P. Elliot and his partner Ithiel Town for a fireproof building for the Patent Office and recommended it to the President.[16] They also approved plans for the Treasury Department as submitted by Mr. Elliot, and they later passed a statute in 1837 to pay William P. Elliot for both sets of plans.[17]

Robert Mills

President Jackson undoubtedly never forgave William P. Elliot for persuading the Senate to publish the results of the Craig hearings. He also had a favorite architect, Robert Mills, for whom he had obtained a temporary job as Patent Office draftsman in May 1835. Mr. Mills designed modifications for Jackson's home, the Hermitage, after it was burned. We can suppose that Jackson very much wished to replace Elliot with Mills as the architect supervising the construction of the Treasury Building and the Patent Office Building, but this would be difficult with Elliot as the winning architect, according to Congress, for both buildings. Jackson, at the request of Amos Kendall, the Postmaster General, had asked Mills to modify his Treasury Building plan to add sufficient room for the General Post Office. Obviously, he did not ask Elliot to modify his plan. Jackson chose a site for the Treasury Building, as he was allowed by Congress, different from that originally proposed. Then he substituted Mills' plan at the

[15]See 24th Congress 1st Session 1836, Statute I, Chapter 353 Section 7; [Cam91a], [Har91], [Eva19], and [Edi51a].

[16]See Gorham Parks and others to President Jackson 7/4/1836 and John Ruggles to Andrew Jackson 7/4/1836, [Fen92, pp. 460–61].

[17]See Levi Lincoln and others to President Jackson 7/4/1836 and [Fen92, p. 460].

new site for Elliot's already accepted plan. With one of the two buildings designed by Mills, he then made Mills the constructing architect for both, cutting out Elliot as he undoubtedly intended to do from the beginning. The L'Enfant plan for Washington had intended the view along Pennsylvania Avenue from the Capitol to the White House to be unobstructed, but Jackson, according to tradition, in defining the new version of the Treasury Building to be designed by Mills, thrust his cane into the ground at the south end of the proposed building, saying it should stand there. The obstructed view was objected to before the building was finished. It is arguable that the view was obstructed to spite Mr. Elliot.[18]

In later years, Mr. Mills would deny that he had ever seen Mr. Elliot's plans for the inside of the Patent Office, and Mills claimed that he had made up his own set of plans.[19] Those who knew better pointed out that the plans were, for a long time, on public display in the Patent Office[20] and that Mr. Elliot had been paid $300 by Congress for them. Mr. Elliot had met this accusation before, and when he asked Mr. Ruggles for a letter specifying what happened, Ruggles replied that the Patent Office Building had been built according to Elliot's plans, with two or three minor departures, and that he questioned whether any of the departures were actually improvements.[21] A similar letter was received from Henry Ellsworth.[22]

Andrew Jackson

The Patent Office was to be built in the area bounded by F, G, Seventh and Ninth streets in the northwest quadrant of Washington. The first section of the building to be built was the F Street side. Major Pierre L'Enfant, who designed the original layout of Washington, had reserved this area as the site

[18]See papers of Pasquale Joseph and Bianca Morse Federico, Smithsonian National Museum of American History Library, [Cla40], and [Bry16, pp. 241–48].

[19]See [Edi51b] and [Edi51c].

[20]See [Edi51a].

[21]See John Ruggles to William Parker Elliot 2/27/1841 in [Fen92, pp. 462–3].

[22]See Henry Leavitt Ellsworth to William Parker Elliot 12/14/1840 in [Fen92, p. 463].

for a pantheon or national church for no particular denomination, but to serve as a home for national monuments, funeral orations, etc.[23] But this was to be a society that had no national religion. The people running the government decided they did not need a building to honor the nation's heroes, and so they decided to build the Patent Office on the site. Looking years into the future, this was to be the site of the Patent Office model rooms in the Heroic Age of American Invention, when people flocked to the Patent Office to see the inventions which inventors were turning out in ever-increasing numbers. At its peak, over 100,000 visitors a year flocked through the Model Rooms to see the inventions and portraits of the great inventors who had changed the world the visitors lived in. The government found that it had quite inadvertently built a building to honor the nation's heroes. But the patent laws were to change, models were no longer to be required, and thus the Model Rooms were no longer required. The Patent Office was to move into a more modern building. The Old Patent Office, as it was by then called, was turned into an art museum, specifically a gallery filled with portraits of the most noteworthy citizens of the nation. Quite inadvertently, the government again fulfilled Major L'Enfant's plans for his reserved area. Perhaps the good major knew what had to be on that site.

When work was to begin on the New Patent Office Building, as it was then called, in July 1836, the land had been occupied for at least 30 years by a squatter, John Orr, a truck gardener who had an orchard extending as far as Mt. Vernon Square. Mr. Orr used to carry a gun to his orchard to scare off boys who were stealing fruit, until one day when the boys stole his gun and discovered that he did not load it. He also found that someone had been milking his cow and stealing the milk, and he decided that it must have been a blacksnake. Eventually, most of the land he was occupying was sold, and his orchard was cut down. This misfortune so depressed him that he soon died and was buried in a pauper's grave. His widow "Jimmy" Orr was occupying a cabin squarely in the middle of the proposed site for the New Patent Office. President Jackson proposed to build the Patent Office off center in the lot so as not to disturb the cabin. The President left the decision of where to build to Commissioner Ellsworth, who then decided to build the building squarely in the center of the F Street side of the land.[24] The workmen who were to build the Patent Office built the widow a small cabin on a far corner of the land, to which they removed her belongings. She died soon thereafter and was buried near Mr. Orr.[25]

The construction of a large fireproof Patent Office was well underway by the middle of July 1836.

[23] See [Cla40].

[24] See William Parker Elliot Diary, 7/8 and 7/11/1836 in [Fen92, p. 467].

[25] See [Pro31] and [Pro35].

Chapter 17

. . . and also the Worst of Years

Henry Ellsworth was appointed the first Commissioner of Patents, and he was confirmed on July 4, 1836.[1] By July 8, he had designed a seal as required for the new Patent Office, and it was approved by the President, with the addition of the words "Seal of the Patent Office."[2]

Clerks under the old law were immediately reappointed under the new law. Joseph W. Hand was appointed Chief Clerk on July 18 and Henry Stone was appointed draftsman on July 21. There was some difficulty with appropriations to pay the employees, since the appropriations were under the old law, which had been repealed, and it was necessary to carry the old appropriations forward to the new law in order to pay anyone.[3]

By the middle of November,[4] 308 applications had been filed under the new law, plus 130 held over unissued from the old law. Only 90 of the 438 cases had not been disposed of by that time. The need for new plates for the engraved part of the issued patent and the requirement for engraving a new seal for the Patent Office had delayed issuance of patents.

Charles M. Keller, the examining clerk, the first patent examiner in the United States and thus the first in the world, was already overworked. Here was one man, virtually self-educated, with the duty to examine patent applications

[1]See John Forsyth to Henry Leavitt Ellsworth 7/6/1836 National Archives M40 26:364.

[2]See John Forsyth to Henry Leavitt Ellsworth 7/11/1836 National Archives M40 26:370.

[3]See Henry Leavitt Ellsworth to John Forsyth 7/30/1836 National Archives Record Group 59 entry 246 Ltrs Rcd Vol 1.

[4]See Henry Leavitt Ellsworth to John Forsyth 11/23/1836 National Archives Record Group 59 entry 246 Ltrs Rcd Vol 1.

in every branch of technology that came before the Patent Office. His genius is shown, not by the fact that he did it well, but by the fact that he could do it at all. But, wrote Ellsworth, by November he needed help. Ellsworth suggested that a second examiner should be appointed. He said that the office of examiner was one of great importance and high responsibility, requiring industry, skill and experience. Few persons could perform the duties of the office, and there were 90 cases pending for which immediate examination was urged by applicants. It was necessary to consult seven or eight thousand models and drawings, many caveats, and European books connected with the subject, frequently written in foreign languages.

The models of the Patent Office, including several thousand stored in the garret of the Post Office, had been classified as well as space would allow in order to facilitate this examination. The $1,500 appropriation for the library had been partially expended on necessary books and an additional $500 was requested. By this time, the Patent Fund contained $160,000 in surplus funds. Much of this surplus was to be spent for the new building, now well under construction. If these were still the best of times, the worst of times were not far behind. William Tell Steiger, already employed as a draftsman in the General Land Office, had a lucrative side business in doing patent drawings for applicants. He was a good friend of Henry Ellsworth and lived directly across the street from the Patent Office. He frequently had to see what had already been designed before he could do proper drawings for new inventions, and he obtained this information mostly from the Patent Office records. Sometime on or before December 14, 1836, he borrowed a book from the Patent Office library to use at his home across the street in doing a drawing he had been commissioned to do. It was a violation of Patent Office rules to remove books or papers from the office, but it was probably vaguely tolerated in Steiger's case. He took the book The Repertory of the Arts and Manufactures for 1797 across the street to his home.

Most of the space in Blodgett's Hotel was occupied by the headquarters of the General Post Office, as the Post Office Department was called, and by the Washington City Post Office. The western three-fifths of the building (the old portion) was occupied by the General Post Office, while the first floor of the eastern two-fifths (the new portion) was occupied by the City Post Office. The second and third floors above the City Post Office, as well as most of the garret or attic of the entire building, were occupied by the Patent Office. The cellar running the full width of the building was dark and damp, with only wooden shutters in the windows and thus subject to drafts. The floor was of dirt or, in wet weather, of ankle-deep mud. At one end of the building, the window was not even closed by a shutter and was open to prowlers. There were three areas used for storage of wood for winter fuel. The General Post Office had an area under its portion piled from floor to ceiling with wood. Under the eastern portion, there were two firewood storerooms, one for the City Post Office and one for the Patent Office. It was the practice of the messengers of the City Post Office and the Patent Office to store ashes from their fires in the cellar. The Post Office ashes were piled on the ground in the corridor, and the Patent Office ashes were kept in a wooden box in the corner of the fuel room. The messengers had been warned against storing the ashes, but there was no trash pickup in those days, and they had nothing else to do with them. The ashes from General Post Office were dumped in the street, along with the piles of garbage dumped there by the rest of the community.

William T. Steiger

There was a fire-engine house at the northeast corner of Blodgett's Hotel, complete with a fire engine purchased by act of Congress 16 years earlier. The engine was a forcing pump with 1,000 feet of riveted leather hose. When it was built, a volunteer fire company was formed to man the engine, but the members became discouraged and disbanded. A few of the employees of the City Post Office worked there until about 2:30 a.m. on Thursday morning, December 15, 1836, until they had delivered the southbound mail to the driver who would deliver it to the steamboat. After that, the messenger, Samuel Crown, went to sleep in the postmaster's office, the watchman, James Summers, went to sleep in his own room, and a Post Office clerk, Cornelius Cox, slept in a room adjoining the letter room. About 3 a.m. Mr. Crown was awakened by suffocating smoke. He examined the fireplace and could find no fire and then went to wake Mr. Summers. Mr. Crown quickly explored the windows in the cellar and found smoke coming out of the southeast end of the building. He next went and woke Mr. Cox, then ran down the street in his night clothes, yelling fire, to the home of another Post Office clerk, James A. Kennedy, and roused him. He then returned to the Post Office to don his pants.

As he ran down the street yelling fire, Mr. Crown probably awoke someone sleeping in the front room of William T. Steiger's home across the street from the Patent Office. Something awakened a friend of Mr. Steiger, who notified Steiger that the Patent Office was on fire. Then, as Mr. Cox knocked on James

Kennedy's door, his knocking probably awoke Henry Bishop, the Patent Office messenger, who lived in the vicinity. Mr. Bishop sent his eldest son to notify Mr. Ellsworth, a few blocks away, and went himself to the Patent Office to get in. He had his keys, but the main door was blocked and he could not get in. He then tried to enter through the City Post Office, but could not make it through the smoke. Steiger went also to notify Mr. Ellsworth but was delayed because Mr. Ellsworth had recently moved from a boarding house to his own house on C Street. Steiger got there just as Ellsworth was preparing to leave. They went together to the Patent Office and tried with Mr. Bishop to get into the office. It was impossible. They tried to find a ladder to enter the office through a window, but there was no ladder closer than the Treasury Department.

The employees of the Post Office Department at the other end of the building quickly began removing all of the departmental records, and virtually nothing of any importance was lost.

Very little was done initially to extinguish the fire. The engine room next door to Blodgett's Hotel was opened, but the leather hose had disintegrated to the point of being useless. The members of the former volunteer fire company had known this for years. John Ruggles, who boarded nearby at 7th Street and Pennsylvania Avenue, was quickly on hand. He had the engine pulled out of the engine house and tried, but that was found useless. He then formed a bucket brigade, the first useful attempt to put out the fire. His opinion was that if an engine had been available for use within 15 or 20 minutes of his arrival, the fire could have been extinguished with little difficulty. John C. Callan, a local druggist, went over to 14th Street and with great difficulty obtained the engine there and brought it back over to Blodgett's Hotel on 8th Street. By the time the engine arrived, flames were coming through the first-floor windows. Water was poured onto the fire for a while, but the limited supply available was soon exhausted, and flames took over for a final time.[5] Former President John Quincy Adams, now in Congress, heard the alarm bells and went by before sunrise to witness the end of the fire.[6]

Meanwhile, William T. Steiger had returned to his rented home across the street to attempt to save his dwelling. His wife Maria was nine months pregnant. Some of his friends had gathered at his home, and they took Maria and a few small belongings to their home at a safer location. Mr. Steiger spent his time on the roof and in the attic, trying to keep the blankets on the

[5]See Sen Doc 58 24th Cong, 2d Sess 1/9/1837; HR Doc 134 24th Cong 2d Sess 1/20/1837; Sen Doc 215 24th Cong 2d Sess 3/2/1837 and [Fed37].

[6]See [Ada77].

roof wet. Glass in the attic windows had melted and cracked, the paint on the house was blistered, and the blankets were badly scorched in places.

He was certain that his home would burn, but it did not. It was, he said, a fortunate circumstance that the wind blew from the west and not from the north, for otherwise he could not have saved the home. About 24 hours later, on Friday morning, Maria gave birth to a baby girl. The parents were thankful that the birth was not 24 hours earlier.[7] When Steiger was later ready to return the book he had borrowed, the library was gone. Children played in the ashes, finding tiny metal pieces, gears and wheels, forever disassociated from the inventions they had helped to model.[8]

Steiger's Sketch of Burned-out Blodgett's Hotel

In fact, every paper, book and model in the Patent Office was destroyed. About 10,000 patents had been issued in 46 years, and they were all destroyed. The models in the garret, which had been so carefully classified, were gone. The original bound volume of full-color patent drawings made by Robert Fulton was burned. Everything—gone. Everything that Dr. Thornton had prevented the British from burning in 1814 was gone. Figuratively and perhaps literally, the musical instrument that Dr. Thornton had saved in 1814 had been burned in 1836. Most of the early industrial history of the nation had been lost to posterity.

Congress, of course, set out immediately to investigate the fire. This was not, as patent people might suspect, because of the loss of 46 years of patents. When the Treasury Building had burned down three years earlier, Richard H. White was suspected and charged with burning the building to destroy

[7]See WTS to AS 12/15/1836 and William Tell Steiger to Wiliam Shriver 12/19/1836, manuscript papers of Shriver family, Maryland Historical Society, Baltimore, Maryland.

[8]See Maria Steiger to Elizabeth Shriver 12/22/1836, manuscript papers of Shriver family, Maryland Historical Society, Baltimore, Maryland.

fraudulent pension papers. The Post Office Department was currently under investigation for awarding dishonest mail contracts, and Congress was having difficulty obtaining the proper records from the Post Office. There was immediate suspicion that Blodgett's Hotel had been burned down to destroy the records of those contracts. The principal problem with that suspicion was that the officials of the Post Office Department had managed to save virtually all of their records from the fire. It was the Patent Office that lost everything, but the Patent Office was not under investigation. Extensive testimony was taken from everyone who might have had anything to say, including a newspaper carrier who was on the street that morning but did not even go to the fire. William T. Steiger provided a drawing of the burned Patent Office which was published in one of the official records of Congress. The tentative conclusion was that someone had stored ashes containing live embers in the basement, and that they set fire to the wood in the basement. There was no finding of which office the ashes came from.

By January 9, 1837, Mr. Ruggles had a bill before the Senate to do what was possible to restore the burned patent records. All patentees were to be requested to return their patents to the Patent Office for copying, and clerks of all courts to which certified copies had been sent were to return them for copying. The Commissioner and two other persons were to be temporary commissioners to decide which burned models were most valuable and interesting, and the Commissioner was to have them rebuilt from available records. No burned patent was to be valid and enforceable until restored. As many temporary clerks as needed for the work were to be hired. In addition, a second examiner was to be hired.

Immediately after Blodgett's Hotel was burned, the Patent Office was temporarily conducted out of Commissioner Ellsworth's dwelling house on C Street.[9]

[9]See Henry Leavitt Ellsworth to John Forsyth 2/11/1840 National Archives M687 10:168–86.

Chapter 18

The New Order in Charge in the Patent Office

On October 31, 1836, Dr. Thomas P. Jones resigned his clerkship in the State Department, and he then opened his office as a patent agent in Washington. One of his first clients was George Escol Sellers, who had invented a furnace to burn anthracite coal.[1] Although Dr. Jones had, for at least eight years, published a report of the patents issued each month, the issue of The *Journal of the Franklin Institute* for January 1837 was without any such report, since the papers which he was preparing for publication were burned in the desk that he still maintained in the Patent Office. About March 1837, Dr. Jones was appointed the second patent examiner.[2]

During the wait for the completion of new Patent Office Building, some suitable accommodations were needed for the Patent Office personnel, growing in number. In September 1835, there were six employees of the Patent Office. By September 1837, there were 25 employees, of whom 14 had been hired to help restore the burned patents. The city authorities of Washington City offered the use, without rent, of fireproof rooms in the new city hall, parts of which were still under construction (and which still stands as the U.S. District Court for D.C.) The only charge made was for structural alterations necessary to accommodate the Patent Office in the building.

[1]See [Sin74] and Patent 170 of 4/20/1837.

[2]See papers of Pasquale Joseph and Bianca Morse Federico, Smithsonian National Museum of American History Library.

Washington City Hall (When Completed)

On January 1, 1838, Commissioner Ellsworth issued the Annual Report of the Patent Office for 1837.[3] Although there had been intermittent annual reports before this time, the Annual Report for 1837 began a long unbroken chain of annual reports first required by section 14 of the Act of March 3, 1837, making it the duty of the Commissioner to lay before Congress in the month of January annually, a detailed statement of expenditures, payments and patents issued and expired, "together with such other information of the state and condition of the Patent Office as may be useful to the Congress or to the public." These annual reports were published by one or both houses of Congress, and, until the advent of the *Official Gazette of the U.S. Patent Office* in the 1870s, were in great demand by libraries, manufacturers, farmers, and inventors. The Annual Report included an agricultural report as a separate volume from 1849 through 1861. During some administrations, when the needs of the Patent Office were subordinated to those of the executive department to which it was attached, these reports were a frequent embarrassment to the cabinet officer in charge, whose shortcomings were loudly and publicly proclaimed by a Commissioner who not only had a right to report directly to Congress, but had a personal duty imposed by statute to provide Congress with such information as might be useful to Congress and the public, however embarrassing the administration might find it.

There was an immediate reduction in the number of patents granted under the new law. In 1835, the last full year under the old law, 757 patents were issued. In 1837, the first full year under the new law, 435 patents were issued. The Commissioner wrote that about one-third of applications being filed under the new law were rejected. That would make about 650 applications, all of which would have been patented under the old law. Perhaps the rest of the

[3]See [otCoP37].

difference can be accounted for among those whose inventions were not new, and they knew it when they filed under the old law. Under the new law, these applications would not be filed.

In the effort to restore the burned patents, of which there were over 10,000, the Patent Office sent a letter to each inventor at the residence given at the time the patent was issued. Many inventors, of course, had changed their residences or died since their patents were issued, but many had been reached. At the conclusion of 1837, 2,000 patents had been restored. Eventually, 2,845 patents were restored. In later years, most of these patents were to become the X series of patents, each being given a number beginning with X to distinguish it from the new series of patents beginning in July 1836 with Patent No. 1 to John Ruggles. Of course, a few of the burned patents were in the new series, which began about five months before the fire.

There was at this time a rapid increase in inventions for improved agricultural implements to save the labor and improve the productivity of farmers. The Commissioner noted that inventors were optimistic, and probably not without reason, that the time was not far distant when plowing machines would be driven by steam. It was for continued belief in this statement that Mr. Ellsworth's last will was contested on the grounds that he must have been insane.[4]

But, wrote Ellsworth in the Annual Report for 1837, farmers might get great benefit from the establishment of a regular system for the selection and distribution of grains and seeds of the choicest varieties for agricultural purposes. The Patent Office was crowded with men of enterprise, who, when they brought in their improved agricultural implements, were eager to communicate their knowledge of new and valuable varieties of seeds and plants. Many had given wonderful new varieties of seeds to Commissioner Ellsworth to distribute where he thought they would do the most good. Commissioner Ellsworth thought there should be a national depository of articles of that kind, from which they could be distributed to farmers in every part of the United States. He suggested that naval officers and diplomats in other countries in other parts of the world might willingly make collections of improved seeds and plants if they had a place to send them for the benefit of the nation. Selection of strains of Indian corn, for use in the Northern states in particular, could increase the crop by one-third with no additional labor. Mr. Ellsworth proposed to, and did, continue to distribute improved seeds in small packages to farmers throughout the country for their test, evaluation, and possible future large-scale use. The initial cost was low. The ultimate result, to be seen about

[4]See National Cyclopedia of American Biography 7:516.

twenty-five years later, was the splitting off of the Department of Agriculture from the Patent Office, where it had outgrown its parent's home.

By the Annual Report for 1838, a problem had been observed with the appeal procedure from adverse decisions of the patent examiners. A board of examiners was authorized in the Act of July 4, 1836, to be composed of three disinterested persons appointed as arbitrators by the Secretary of State, to hear any appeal from the Patent Office. This board was to be paid $25 per appeal. However, the applicants were introducing extensive additional evidence on appeal which had not been before the Patent Office on the initial examination, and the arbitrators on the board of examiners, who differed from case to case, frequently had no judicial experience and no knowledge of how to prevent being overwhelmed with time-consuming proceedings in a case which they were expected to decide for a total of $25, divided between them. It was becoming hard to find willing arbitrators. The Commissioner suggested that the appeals should be to the Chief Justice of the District of Columbia, who would have authority to determine the matter in chambers if he wished, or to allow introduction of evidence if he wished.

A problem was also being observed with the limited space available for the Patent Office in the City Hall. As more and more models arrived with applications or were reconstructed to replace burned models by the Patent Office, there was no space to display them. The Commissioner protested delays which were apparently occurring in the construction of the new Patent Office Building. He also observed that work in the Patent Office was being delayed by lack of examiners. Already the Patent Office had two examiners, and the Commissioner asked for two assistant examiners to help them. There were over a hundred patent applications on hand which had not been reached by the present examiners. This number would delay issuance of patents for several months after filing the applications. Meanwhile, applicants were arriving in the city from distant parts of the country, expecting to file an application and take home a patent three or four days later.

By the Annual Report for 1838, Congress had asked the Patent Office to estimate the cost of collecting agricultural statistics for the nation. Ellsworth rejoiced in this inquiry, noting that a failure of the corn crop in the Middle and Southern states in the previous season had persuaded fearful farmers to make forced sales of their livestock, fearing that they could not feed them during the winter. Had there been good national crop statistics collected, they would have been aware of the unsurpassed crop of Indian corn on the western side of the Appalachian mountains, leading to very low prices on corn which could be shipped down the Missouri River. Meanwhile, the newly introduced "Baden" corn in the Mississippi valley promised to increase the crop of Indian corn by

50 percent. Good crops of corn were being raised on the borders of Canada, and strains of Siberian and Italian spring wheat had been distributed by the Patent Office.[5]

By December 1838, William P. Elliot was at his apparent hobby again, reporting to Secretary of State John Forsyth that Dr. Jones had been preparing patent applications for inventors while employed by the Patent Office. Forsyth wrote to Ellsworth that he was informed that Ellsworth had known of the matter for two months and required him to furnish an explanation.[6] Ellsworth replied that Mr. Elliot had confidentially informed him weeks earlier of his intention to file charges against Jones soon but said he was not ready to give his information yet. Ellsworth then asked Dr. Jones if he had prepared a specification for Asahel Collins since his appointment and was assured by Jones that he had not. Only a week before Forsyth's letter, Ellsworth was told by Elliot and Collins that a paper in Dr. Jones' handwriting, which was of record in a case between Collins and a Mr. Demphel, showed that Jones prepared Collins' specification. Ellsworth then offered to start an investigation immediately if they would leave this paper with him. Collins refused to do so while his case was pending.[7] In the reply to Forsyth was a letter from Jones to Ellsworth, concurred in by Collins, stating that he drew the specification for Collins, having been his agent prior to his appointment as patent examiner.[8] Another letter was included from Jones to Ellsworth in which he said that he had not undertaken any business from Collins directly or indirectly, had not received a cent from him, and that there was not then anything depending between them for which he was to derive any benefit.[9] Secretary Forsyth replied that President Van Buren had seen the letters and had required the case to be investigated under oath, both as to Dr. Jones and as to the delay of Mr. Ellsworth. Dr. Jones was to be required to show that he had not rendered any services for Collins since he was appointed for which he, at the time, expected to receive compensation. Mr. Elliot and Mr. Collins were to be requested to state their charges under oath.[10]

Ellsworth replied with details to show that he had no information on which to start an investigation until a week prior to Forsyth's first letter, and that

---

[5]See [otCoP38].

[6]See John Forsyth to Henry Leavitt Ellsworth 12/13/1838 National Archives M687 17:546.

[7]See Henry Leavitt Ellsworth to John Forsyth 12/14/1838 National Archives M687 17:548–54.

[8]See Thomas P. Jones to Henry Leavitt Ellsworth 12/8/1838 Note 7.

[9]See Thomas P. Jones to Henry Leavitt Ellsworth 12/13/1838 Note 7.

[10]See John Forsyth to Henry Leavitt Ellsworth 12/17/1838 National Archives M40 28:94–95.

he began his investigation almost immediately.[11] William P. Elliot also wrote to Forsyth, indicating that Mr. Ellsworth offered to investigate the charges immediately when they were called to his attention, but Mr. Collins would not furnish him with the evidence. Elliot also wrote that at some later time, charges would be brought, "however painful it might be to the person whose duty it would be to prefer them." Apparently, Mr. Elliot would have the Secretary believe that he did not like bringing charges. Sometime prior to mid-January 1839, Thomas P. Jones wrote one or more letters to the President, resigning his offices as patent examiner and member of the board for restoring the patent models, but saying he was leaving office with clean hands.[12] Commissioner Ellsworth was totally exonerated, and Dr. Jones was replaced as examiner by Dr. Thomas W. Donovan of Maryland.

The Annual Report for 1839, dated January 1, 1840, said that the new Patent Office Building was within a very few weeks of completion, and was soon to provide accommodations for the office staff, room for the models and specimens, and facilities for collecting and distributing seeds. There were about 800 applications filed in 1839, of which 400 were rejected on examination, with no appeal being taken from any rejection.

The Annual Report for 1840 indicated that the new building was occupied, with adequate room for displaying models. The Patent Office distributed 30,000 packages of seeds that year.

In February 1840, Mr. Ellsworth learned that Henry Bishop, the messenger, had been stealing franked envelopes and using them for private correspondence. He was induced to resign.[13] A few days later, Bishop filed charges against Ellsworth for abuse of franking privilege, neglecting his duties, stealing office stationery, etc.[14] Ellsworth immediately wrote that he was always the first in and last out of the office and the rest of the charges were also untrue. Nothing resulted from the charges.

In 1841, the Secretary of State approved the use of the temporarily empty large hall in the new Patent Office Building by the National Institute, an association incorporated for twenty years to have custody of the personal effects of James Smithson, founder of what was later to be the Smithsonian Institution, of the collections made by the Wilkes exploring expedition, and donations from many societies and individuals. In addition to the collections of the National Institute and the patent models,

[11]See Henry Leavitt Ellsworth to John Forsyth 12/20/1838 National Archives M687 17:556–63.

[12]See William P. Elliot to John Forsyth 12/24/1838 National Archives M687 17:565–67.

[13]See Henry Bishop to Henry Leavitt Ellsworth 2/4/1840 and others National Archives M687 2:384–91.

[14]See Henry Leavitt Ellsworth file National Archives M687 10:160–292.

many costly or invaluable objects belonging to the State Department were to be displayed in the National Gallery. An appropriation was requested for watchmen to guard the building and for fencing the public square on which the building stood to prevent valuable objects from being stolen.

Mr. Ellsworth noted that agricultural chemistry was progressing at a good pace. It was becoming possible to extract oil for oil lamps from lard and also from corn, whereas previously only whale oil or possibly olive oil burned well in an oil lamp.[15]

Charles G. Page

Toward the end of 1841, Commissioner Ellsworth wrote a private letter to Secretary of State Daniel Webster,[16] pointing out that the two examiners, Keller and Donovan, disliked each other so much that they would communicate with each other only in writing. Mr. Ellsworth said that this was extremely embarrassing to the Patent Office, that the only alternative was a change of examiners, and that he could make this statement with no political motive. Soon afterwards, Thomas W. Donovan left the Patent Office on April 1, 1842, or shortly before and was replaced by Charles G. Page. There were by this time also two assistant examiners, Henry Stone and William P.N. Fitzgerald.

In the Annual Report for 1843, Commissioner Ellsworth included for the first time the claims of all patents issued during the year. At that time, no periodical in the country published the claims of all patents, and half the patent applications filed were rejected because the applicants could not know what had been done previously in all cases. Also, for the first time, the Commissioner had the examiners prepare individual reports of the progress of the arts that they examined for inclusion in the Annual Report of the Patent Office.

Ellsworth held forth in this report on the reduction in cost of commonly purchased items. Shirt cloth which had cost 62 cents a yard thirty years earlier now cost eleven cents a yard. Hooks and eyes thirty years earlier cost $1.50 a

[15] See [otCoP41].

[16] See Henry Leavitt Ellsworth to Daniel Webster 12/7/1841, papers of Daniel Webster (Microfilm edition Library of Congress microfilm 13,857 reel 3).

gross, now cost fifteen cents a gross. Horseshoes, formerly turned out one at a time by blacksmiths, were now sold ready made at five cents a pound.

Norbert Rillieux

The report carried a reprint from the *New Orleans Bee* of the invention by Norbert Rillieux (1806–1894) of an apparatus for the evaporation of molasses to produce sugar while heating the molasses entirely by steam. A number of vacuum pans were used in series to evaporate the water with the use of only one third of the fuel used in the usual method. Mr. Rillieux called this his triple-effect evaporating apparatus. R.J. Packwood, who made the report, said that he found the apparatus was arranged with the greatest of care and very durable. It worked without any accident and was ready to be used again during the next season. Mr. Rillieux was black, or at least of mixed blood, and was educated in France. His father Vincent Rillieux was a French engineer and his mother Constance Vivant was a light-skinned black woman. About 1861, when restrictions on free blacks were increased in Louisiana, he returned to France, to spend the rest of his long life. In his later life, he worked to decipher Egyptian hieroglyphics.[17]

Mr. Keller, still one of the examiners in 1843, noted that although the models of early inventions had been destroyed in the fire of 1836, it had been his privilege "in boyhood and early manhood, to study the models of those beautiful specimens of American ingenuity, many of which yet live in my recollection." He also wrote: "Since the invention of Oliver Evans, which has given him enviable reputation wherever flour is used, no marked invention in mills has been made, although many minor improvements have been patented and introduced in the manner of dressing mill stones." The other examiner, Charles G. Page, also made a report, not quoted here, it having little of broad historical interest.

Mr. Ellsworth wrote one sentence in the 1843 report which has been misunderstood and misquoted ever since. He wrote: "The advancement of the arts, from year to year, taxes our credulity, and seems to presage the arrival of that period when human improvement must end." The statement which is usually falsely attributed to some Commissioner or another, based upon this, is that "Everything that can be invented has been invented." No Commissioner has

[17]See [HD94] and[Top71].

ever said this, and probably no Commissioner has ever thought it. In his 1988 book, *Victory without War*, Richard Nixon attributed the latter statement as of 1899 to Commissioner Charles H. Duell, who also never said it.[18] In his last Annual Report, the one for 1844, Commissioner Ellsworth wrote that 502 patents issued during the year, up from 435 during 1837, the first full year under the new law. He noted that the inventory of models was expanding so fast that there was no room to properly classify them, that the models of rejected applications were not exhibited at all for lack of sufficient room, and that the beautiful collection of curiosities forming the National Gallery, while occupying necessary room, were too important to be crowded out. He recommended adding the contemplated west wing to the original building. The law had given the power to a board of commissioners, consisting of the Secretary of State, the Commissioner of Patents, and the Solicitor of the Treasury, to extend patents issued almost 14 years ago and about to expire, for an additional seven years, provided the inventors have not yet been able to make the invention profitable. Mr. Ellsworth noted that none of the patents for which extensions were requested had been examined for novelty under the old law under which they issued, and that the board of commissioners would not extend a patent if it were not shown to be novel. The authority to extend patents was given under the Act of July 4, 1836, and the board had extended seven patents since that act. Of course, Congress had always had, and has to this day, the power to extend any patent for what seems sufficient reason.

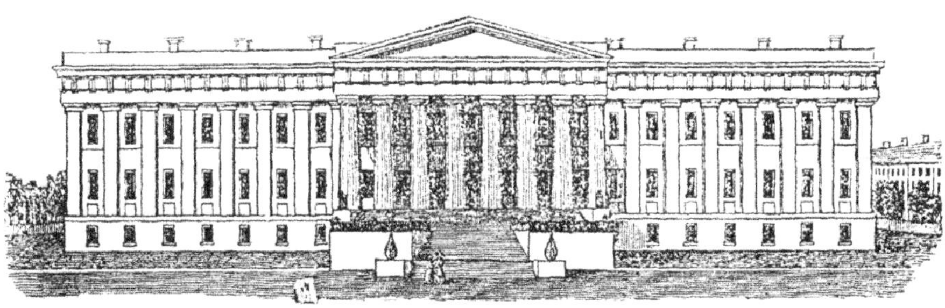

New Patent Office in 1846

Charles M. Keller spent several years studying law at night after office hours, and in May 1845, he started a long tradition among patent examiners by leaving the Patent Office to practice patent law. He initially established an office in Washington, but soon established a successful patent law practice in New York City.

[18] See [Jef40] and [Sas89].

Chapter 19

What Hath God Wrought

The steed called Lightning (say the Fates)
Was tamed in the United States
'Twas Franklin's hand that caught the horse
'Twas harnessed by Professor Morse.[1]

Indeed, it was Professor Samuel Finley Breese Morse (1791–1872) who first commanded the lightning to labor. And, yes, they laughed at Professor Morse when he tried to tell them, in the closest analogy that he could find, that he could send messages by lightning. "Canst thou send lightnings, that they may go, and say unto thee, Here we are?"—Job 38: 35 Answer: Yes, thanks to Professor Morse. But we are ahead of our story.

Henry Ellsworth had been a Yale classmate of Samuel F.B. Morse of telegraph fame, and they remained friends in later life. Professor Morse was a portrait painter of some note, a professor of the literature of art at the University of the City of New York, and a founder of the National Academy of Design. He came to technology later in life, having first conceived the idea of a telegraph on a trans-Atlantic voyage in 1832.

Professor Morse exhibited his telegraph in 1837, signed his patent application in 1838 and got patent 1,647 in 1840. Whenever he came to Washington City on business, he visited the Ellsworths, although, as was the custom of the day, he stayed in one of the many Washington boarding houses. The 1837 exhibition of the telegraph was met with an immediate and overwhelming lack of interest. As early as 1838, Morse attempted to secure the aid of Congress in constructing a telegraph line from Washington to Baltimore to

[1] See [Edi55b].

test the abilities of his invention. He did not achieve early success in this endeavor. Congressmen belittled the mad professor who, they believed, thought he could send messages via lightning, the only electricity they understood.

Samuel F.B. Morse

In 1842, when both Ellsworth and Morse were in Paris for an international exhibition, Ellsworth wrote back to friends in the United States commenting that if one telegraph instrument were placed in the Capitol and another connected to it in New York City, the people of New York City would know the results of a vote in Congress before it was known at the White House at the other end of Pennsylvania Avenue. This was at a time when the fastest known way to send a message over a long distance was to give it to a messenger on a fast horse and send the messenger on his way.

In 1843, Morse had another bill pending in Congress to appropriate $30,000 to construct an experimental telegraph line between Washington and Baltimore. Many Congressmen found the bill to be an object of ridicule, and they proposed many amendments to the bill to show their scorn. One amendment proposed that half the money be used not to test electrical magnetism, but animal magnetism, also then called mesmerism and now called hypnosis. But eventually the bill made its way through the House, which passed it and sent it on to the Senate. On March 3, 1843, Congress was in the last day of its session, and the bill seemed destined to be lost to legislative procrastination.

On the last evening of the session, both Morse and Ellsworth were in the visitors' gallery of the Senate or in the lobby, working for passage of the bill and waiting for something to happen. Finally, as the evening lamps were lit, there were 119 bills ahead of it. One of Morse's friends in the Senate came by and mentioned that he thought it was impossible that the bill would be reached for a vote. Professor Morse was so discouraged that he left the Capitol, bought a ticket on the newly installed steam-driven trains for his home in New York City for use the next day, then went back to his boarding house and paid his bill. This left him with less than one dollar to his name. Then he went to bed, prepared to leave Washington as an abject failure the next morning.

But Henry Ellsworth remained at the Senate, lobbying doggedly for the bill. Finally, five minutes before adjournment, the bill was passed by the Senate. Only one other bill passed afterwards before the Senate ended its session. The President was on hand to sign the bill into law before midnight. Henry Ellsworth explained the success to his family the next morning. Annie Goodrich Ellsworth (1826–1900), the Commissioner's seventeen-year-old daughter, according to family tradition had a teenage crush on Samuel Morse, who was a fifty-two-year-old widower. Professor Morse was polite to Annie, which seems to have been all she required to maintain her interest. When Annie found out the good news from her father, she asked for and received her father's permission to go over to Mr. Morse's boarding house and tell him the good news.

Annie Ellsworth

Annie was incidentally a part-time employee of the Patent Office. It was the custom of the day for local women to be hired to copy papers out in longhand. In the days before copying machines or even typewriters, this was the only way to get a copy of a patent from the Patent Office. Annie copied some 13,000 words at 10 cents per 100 words in 1843.

Professor Morse was in the dining room at his boarding house, having breakfast before returning to New York City, when he was told that there was a young lady waiting to see him in the parlor. In the parlor, he found Annie, who asked him if he had heard the good news. He answered that he had not heard any good news recently. She replied that she thought not, that her father had told her she could come and tell him that the Senate passed his bill before its adjournment, and that the President signed it into law before midnight.

Morse thanked her for the news and told her that for bringing him such good news, he would allow her to send the first message from Washington to Baltimore over his new telegraph line when he finished building it. About a year later, when the line was about finished, Morse asked Annie if she had her message ready.

Annie replied that she and her mother had been reading through the Bible, looking for the perfect message, and that they had found, in Numbers 23: 23, her message—"What hath God wrought!" On May 24, 1844, Morse, in the old

Supreme Court chamber of the Capitol, sent her message to Alfred Vail at the Mount Clare depot in Baltimore, inaugurating a new era of communications.[2]

Although the new telegraph connection was between Washington and Baltimore and was openly demonstrated in both cities, the first newspaper account to appear in the National Intelligencer, the leading newspaper in Washington, was on May 27, and that was a copy of an account from a Baltimore paper. Only 16 people showed up in the old Supreme Court chamber to see the Washington end of the demonstration. The newspaper account did not mention Annie's message.[3]

[2]See [Edi55b], [Mor00], [Edi91a], [DotAR07], and [Pri75].

[3]See *National Intelligencer*, 5/27/1844.

Chapter 20

Last Years under the State Department

Commissioner Ellsworth remained in office longer than he had originally intended because he felt obliged to see the Patent Office back on its feet after the fire of 1836. In fact, he remained in office longer than some people thought he would. In 1841, thinking that Ellsworth was about to leave office, Mr. Ruggles, now a lawyer in private practice in Thomaston, Maine, wrote a letter to Secretary of State Daniel Webster applying for the position of Commissioner of Patents.[1] He had guaranteed notes for friends who had defaulted, and he was now near broke. He reminded Mr. Webster of his long career in public service, filling a series of jobs that he had not sought. He said that this was the first such job he had ever sought. He reminded Mr. Webster that he had written the very patent act which governed the operations of the Patent Office. But Mr. Ellsworth did not leave office for another four years, and Mr. Ruggles never became Commissioner of Patents.

When Henry Ellsworth resigned from the Patent Office on April 1, 1845,[2] he became Land Commissioner of the United States in Lafayette, Indiana. There he bought tracts of land and became the largest landowner and farmer in the area. He was a strong advocate of agricultural machinery, and he used what was probably the first mowing machine used on the prairies. When his health failed, he returned to Connecticut in 1857 and died a year later.

---

[1] See John Ruggles to Daniel Webster 4/24/1841, the Papers of Daniel Webster (Microfilm edition Library of Congress microfilm 13,857).

[2] See Henry Leavitt Ellsworth to James K. Polk 4/1/1845 and [Jef40].

Edmund Burke

The second Commissioner of Patents was Edmund Burke (1809–1882), who was a lawyer and newspaper publisher in New Hampshire and from 1839 to 1845 was a member of Congress from New Hampshire. After he lost an election, he was offered the position of Commissioner in the place of Henry Ellsworth. He took office on May 5, 1845.

The first problem of Commissioner Burke was a workload in the Patent Office which exceeded the capacity of the examiners to do the work promptly. He ordered examiners to work ten hours a day until the work was current.[3] However, it seems that increasing the hours of labor did not increase and may have reduced the amount of work done.[4] Soon afterwards, two examiners were added to the staff of the office. It was noted that the business of the office was increasing at the rate of thirty percent per year, and many more new examiners would probably be needed.

Titian R. Peale

One of the new examiners was to be Titian Ramsay Peale (1799–1885), the youngest son of Charles Willson Peale. He was, incidentally, the second son of his father named Titian, the first Titian having died shortly before our Titian was born. Presumably, if you have decided to name all of your children after painters, the name Titian is too good to lose. At the time of his appointment to the Patent Office, he had been in the service of the country, off and on, for about fourteen years.[5] In 1818, while yet a teenager, he entered Stephen H. Long's exploring expedition to the Rocky Mountains as an artist and assistant naturalist to record what they found. This expedition lasted two years, much of the time without tent or rations. He was with the

[3]See Orders of Commissioner Burke 12/7/1846 and 2/28/1848, "Scrap-Book", Department of the Interior, mss in Special collections, Scientific Library, U.S. Patent and Trademark Office, Washington, D.C.

[4]See [Edi48].

[5]See Titian Ramsay Peale to Millard Fillmore 4/29/1852, Peale Family Papers [Kraus Microform], compiled by Smithsonian Institution, Washington, D.C.

Charles Wilkes naval exploring expedition, formed in 1836, until it was completed. He suffered shipwreck in that expedition, but brought back much of the material which was later exhibited in the National Gallery at the Patent Office. He was a member in very good standing of the Franklin Institute. He had a background in museum work from his father's Philadelphia Museum. He had at one time been curator of that museum, which was failing by this time, and he had earlier hoped to be a curator of the Patent Office museum, but that was not to be. He was one of the earliest photographers in the country and was engaged as a daguerreotype-maker in Philadelphia in 1845, when he had a run-in with a gang of thieves associated with Jim Webb, to be heard from shortly.[6]

Titian R. Peale applied for a position as a principal examiner in 1848, but there were only four such positions. One of his earliest letters of recommendation to Commissioner Burke for the position of examiner was from Alexander Dallas Bache, who was a great-grandson of Benjamin Franklin and the Superintendent of the United States Coast Survey. However, Mr. Bache recommended him for a position as an assistant examiner rather than principal examiner, much to Mr. Peale's surprise.[7] Bache quickly corrected that by an additional letter.[8] The amount of influence brought to bear upon the problem of getting a low-level government job is unbelievable by today's standards. Mr. Peale had Professor Joseph Henry of the Smithsonian Institution drop in personally to recommend him to Commissioner Burke and was having all of the influence of the Franklin Institute brought to the problem, but he was afraid that Lieutenant Joseph D. Webster, who was a protege of Gen. Lewis Cass, the 1848 Democratic candidate for President, would have better connections.

While his application was pending, Mr. Peale heard that assistant examiner Thomas G. Clinton had filed charges against Mr. Burke. Indeed, this was so, because Mr. Burke had suspended examiner Clinton pending his removal, and then when he was removed had refused to pay him his salary during the period of suspension. Dr. Clinton then appealed to the President, and Secretary of State James Buchanan sided with Dr. Clinton as to his right to receive pay during the suspension period.[9]

---

[6]See *National Police Gazette,* 12/13/1845.

[7]See Alexander Dallas Bache to Edmund Burke 5/17/1848, Peale Family Papers [Kraus Microform], compiled by Smithsonian Institution, Washington, D.C., VIII-A, fiche 11 of 17, document 005494.

[8]See Alexander Dallas Bache to Edmund Burke 5/21/1848, Peale Family Papers [Kraus Microform], compiled by Smithsonian Institution, Washington, D.C., VIII-A, fiche 11 of 17, document 005494A.

[9]See James Buchanan to Edmund Burke 8/22/1848 and 8/26/1848, manuscript papers of Edmund Burke, Library of Congress, Washington, D.C.

Titian Peale was offered an appointment as assistant examiner, and he initially declined it, even though he had no other prospects.[10] However, on advice of Joseph Henry and others,[11] he did accept the position in August 1848.[12] By May 1849, Peale was settled in and liked his duties, but found his salary insufficient to overcome debts. He had received his first confirmation of a rumor that Commissioner Burke was about to leave by seeing packing boxes lined up in front of his room.[13]

Thomas Ewing

Prior to 1848, patents could issue on any day, but beginning in early 1848 and continuing to date, patents have issued at noon every Tuesday, and only on Tuesday, come fire, flood, war, riot, or national holiday.

On March 11, 1848, Journal of the Franklin Institute editor-for-life Thomas P. Jones gave up his editorship at death. The journal continued despite loss of its first editor.

As far back as 1812, some officials had advocated a Home Department in the Government to handle non-foreign affairs, leaving the State Department to handle foreign affairs. It was to comprise territorial governments, national highways and canals, the General Post Office, the Patent Office, and the Indian Affairs Office.[14] A Home Department was founded by the Act of March 3, 1849. Its first Secretary was Thomas Ewing, and its name was the Interior Department by early April. The 1849 act also appropriated $600,000 to complete the east wing of the Patent Office Building, of which $250,000 was to come from the Patent Fund. The Patent Office was placed within the Home

[10] See Titian R. Peale to John Frazer 6/28/1848, Peale Family Papers [Kraus Microform], compiled by Smithsonian Institution, Washington, D.C., VIII-A, fiche 11 of 17, document 000358.

[11] See Titian R. Peale to John Frazer 6/28/1848, Peale Family Papers [Kraus Microform], compiled by Smithsonian Institution, Washington, D.C., VIII-A, fiche 11 of 17, document 005502.

[12] See Edmund Burke to Titian R. Peale 8/14/1848, Peale Family Papers [Kraus Microform], compiled by Smithsonian Institution, Washington, D.C., VIII-A, fiche 11 of 17, document 005502.

[13] See Titian R. Peale to John Frazer 5/9/1849, Peale Family Papers [Kraus Microform], compiled by Smithsonian Institution, Washington, D.C., VIII-A, fiche 11 of 17, document 000362.

[14] See House of Representatives Document 326 Congress 12:1 and James Monroe to John Gaillard 12/6/1816, American State Papers 2:396–99.

Department, and the Home Department, lacking a building of its own, moved into the Patent Office Building and began elbowing for space.

The conflict between the Patent Office and the Interior Department for space in the building continued for many years and would continue to cause hard feelings among the nation's inventors and their representatives.

The residents of Texas declared independence from Mexico in 1836, and by October the residents, mostly settlers from the United States, had established a permanent government for the Republic of Texas. But Texans were missing something from the old country. The Texas Congress passed, and on January 28, 1839, President Mirabeau Bonapart Lamar signed into law an act securing patent rights to inventors. Texas patents were available only to a Texas citizen or one who had legally declared an intention to become a citizen. The inventor had to come into the Texas Patent Office in person to make his application. The law was administered by the Chief Clerk of the Texas State Department, who was head of the Patent Bureau. A detailed description of the invention was required, including either a drawing or a model of the invention. A report was made by Chief Clerk Nathaniel Amory of the Texas Department of State to David G. Burnet, acting Texas Secretary of State, on November 6, 1839,[15] reporting that six patents had been granted by that time. When the Republic of Texas was admitted to the Union as the State of Texas in 1845, no provision was made for transfer of Texas patents to the United States Government. In the Annual Report for 1846, Commissioner Burke objected to the failure of Congress to provide for the transfer of records and patent rights. But apparently no such provision was ever made.

Dr. John Gorrie was a practicing physician in Apalachicola, Florida, in the late 1840s. He frequently treated fever cases, and in the hot and humid environment of the South, only ice, imported by ship from the North, could reduce fever and give the patient an opportunity to recover. But ice was not always available, and some patients did not recover. Dr. Gorrie learned how to make ice using mechanical energy to drive a compressor. In April 1848, he went upriver from New Orleans to Cincinnati to have an ice machine made according to his invention at the Cincinnati Iron Works. This was in the days long before the use of freon or even ammonia in the compressor, so Dr. Gor-

Dr. John Gorrie

[15]See [Smi32] and [FN36].

rie ran his machine by compressing air. Although it worked, it did not work efficiently. Dr. Gorrie received an English patent for his invention in 1850, and on May 6, 1851, he got U.S. Patent No. 8,080. Dr. Gorrie had sold 75 percent of his invention to finance the construction of the large-scale working model, and when his partner died without a will, Gorrie's invention was left in limbo. When Dr. Gorrie died in 1855, his invention was forgotten for years—but not lost.[16]

[16]See [She82] and [Edi55a].

Chapter 21

The Great Patent Office Jewel Robbery

Actually, there were at least three robberies of jewels from the Patent Office, but perhaps only one great robbery. The National Gallery in the Model Rooms was used to display a number of treasures of the United States government, including the Declaration of Independence, the Constitution, and numerous beribboned treaties with foreign governments. Also included were gifts given to government officials by foreign governments and thus held to belong to the nation. These included jewels and other valuable objects.

On December 20, 1841, not very long after the National Gallery was first opened, a robber using false keys entered a small room containing these treasures at about 1 p.m. He stole an elegant and costly gold snuff box set with diamonds on the lid which had been presented by Emperor Nicholas of Russia to the American ambassador to the court at St. Petersburg, a pearl necklace consisting of 148 pearls, two large separate pearls sent to the President by the Imam of Muscat, and a gold scabbard for a sword. The sword had a diamond studded hilt and was presented by the Spanish Viceroy of Peru to Commodore James Biddle while he was cruising on the Pacific coast during 1830. The sword for the scabbard was left behind to make it possible for the robber to fold up the scabbard for concealment. The total value of the articles stolen was estimated to be $14,000 to $15,000. The room had been opened and the jewels shown only a few minutes before the robbery, and it was liable to be visited at any moment when the robber was doing the deed.[1] Commissioner Ellsworth issued an immediate advertisement offering a reward of $1,000 for recovery of

[1] See *Baltimore Sun* 12/21/1841.

the articles and detection of the robber.[2] A correspondent for the Baltimore Sun said that authorities supposed that the robber was an Englishman, as one such was seen in the building just before the robbery. The correspondent humorously supposed that the robber was a Scotsman, as one was seen just after the robbery sneezing enough to break his neck, supposedly being unwilling to waste the snuff he found in the snuff box.[3]

The account given in the newspapers of the day stated that the stolen articles were recovered on January 10, 1842.[4] Supposedly, police officers Simeon Hays, John Zell and Archibald Ridgely of Baltimore received information which led them to believe that they might find the articles by visiting the cabin of a schooner or brig Mary Bright lying at the head of Smith's dock. Justice of the Peace Henry Snyder issued them a search warrant and accompanied them to the vessel. They found the cabin locked, but it was immediately unlocked by the captain of the vessel. There was a large black trunk in the cabin, which had arrived during the morning in the absence of the captain, and which was received and receipted for by a black deck hand. They had to force the trunk open and found that it contained only the whole of the stolen articles from the Patent Office and a large floor mat in which they were wrapped. The trunk was labeled and directed to the care of James Anson in Richmond, Virginia. The deck hand was able to give a complete and detailed description of the man who left the trunk, and the police thought they should be able to catch him. They did not catch him.

The second occasion when the Patent Office jewels were stolen was the "great" robbery, probably because of the attention the robbery received. During the afternoon of November 8, 1848, according to one theory, one or more robbers entered with the visitors to the National Gallery, then secreted themselves in the attic until everyone had left but the guards. The guards apparently had their minds on other things than guarding that night, because this was the evening after the Presidential election between Zachary Taylor and Lewis Cass, and the guards were following the election results. Despite this being a night with strong moonlight, the choice of time for the robbery showed forethought and good judgment.

According to another theory, the robbers entered the front door by the simple expedient of grasping the key, already in the lock of the front door, with nippers and turning it from the outside. They then proceeded up the stairs and used a skeleton key to open the door of the great hall on the second landing, which they secured behind them with a length of cord.

---

[2] *Ibid.*

[3] See *Baltimore Sun* 12/24/1841.

[4] See *National Intelligencer* 1/12/1842.

Whichever way they began, they rifled a large glass double case in the repository. When concealed alarm bells attached to the gold snuff box, the scabbard and the string of pearls began to ring unexpectedly when the items were removed, they contented themselves with less than the total available loot, and they hurriedly left the great hall by a knotted rope which they had let down to the outside from the window. Apparently, no one but the robbers heard the alarm bells, although they must have rung. The robbers were so successful that the robbery was not discovered until the next morning.

In addition to the items stolen the first time, various gold, silver, and copper medals and coins were stolen, together with a pint bottle of attar of roses. This time, the stolen items were appraised at $20,000. The government offered a $1,500 reward for recovery of the items and conviction of the responsible parties.[5]

George Wilkes (1820–1885), editor of the National Police Gazette of New York City, had access to information which the government had earlier hidden from the public, and he also was very knowledgeable about the activities of the criminal class in the country. In his weekly edition for November 18, he announced his confidence that the criminals were Tom Hand, alias Jacob Shuster, and Jim Webb, acting under instructions from Jim Young, ex-high constable of Philadelphia, a policeman turned bad. The purpose of the robbery was not monetary gain but was to obtain something that they could use to bargain with the government to secure either the release of Ned McGowan from his peril in Philadelphia or a pardon for Charles Webb from the New York state prison.[6]

Mr. Wilkes announced that the Patent Office had been robbed once before of the same jewels, which would have been the 1841 robbery, by Tom Hand, to secure the release from custody on criminal charges of Tom Walker, a former companion in crime of the Webbs. When the government offered a $1,000 reward, Tom Walker announced to certain unnamed officers that if the reward were increased to $1,500 and if he were allowed to slip through his present dangers with the law, he thought he could turn up the swag. When this was done, the authorities were told where to look, and the jewels

George Wilkes

[5]See *National Intelligencer* 11/9/1848.
[6]See *National Police Gazette* 11/18/1848.

turned up in the trunk on board a vessel in Baltimore harbor. Tom Walker was then released from custody.

Having succeeded once, Tom Hand and Jim Webb felt they could use the same scheme again to secure another release. But for this to work, it must be possible for the government to hide its complicity in the scheme. George Wilkes announced their intentions in the public press and also announced that he was totally opposed to such compromise. Rather than allowing the thieves to buy a single day from a ten-year sentence, the jewels should be thrown into the sea. However, he predicted that no harm would come to the jewels even if compromise were refused.

The National Police Gazette had published a long multi-part article on Charles and James Webb in 1845 in its "Lives of the Felons" series.[7] The Webbs were born and trained in London, and after Charles was detected robbing the Post Office in Birmingham, both brothers left hurriedly for New York in 1828. There they became the center of a group of thieves who diligently studied such useful arts as lock-picking, key-making, forgery and counterfeiting. They were thoroughly proficient thieves. One of their associates in this country was Jack Reed. While engaged in a scheme with the Webbs to pass forged bank notes, Jack Reed was captured in Philadelphia by Titian Peale, then a clerk in the United States Bank. He was turned over to Jim Young, from whose custody he "escaped." This was in the early 1840s. Tom Hand, alias Jacob Shuster, was not mentioned in the 1845 article

The robbers, angry because Wilkes had thwarted their scheme by publishing their hidden plans to the world, wrote an anonymous letter to the President on December 15, 1848, berating Wilkes and his co-workers, and even Ned Buntline, who was not associated with them, then promising to return the jewels if the President would take from their paper the government advertisements of the army deserter list.[8] This letter persuaded Edmund Burke to first investigate and then contact Mr. Wilkes.[9] They agreed to transfer the advertisements to another paper to see what would happen. Then the robber wrote a second anonymous letter to the President, asking if the transfer of the advertisements had been done in good faith. Jim Webb was arrested and turned state's evidence for immunity from prosecution. He told where to find the jewels. The jewels were found buried in the cellar of Henry B. Jones in New York City, and the attar of roses in his attic. Jones claimed to have taken them as security on a loan to Tom Hand.

[7]See *National Police Gazette* 12/20/1845.

[8]See *National Police Gazette* 4/21/1849.

[9]See Edmund Burke to Thomas Ewing 5/8/1849, manuscript papers of Thomas Ewing, Library of Congress, Washington, D.C.

The trial of Tom Hand in the U.S. Criminal Court for the District of Columbia began in mid-April 1849 with District Attorney Philip Barton Key (1818–1859), son of Francis Scott Key, prosecuting. Handwriting experts testified that Tom Hand wrote the letters to the President, and other handwriting experts testified that he did not. After several days, the jury agreed that it could not reach a verdict. Seven were for acquittal and five for conviction. A second trial was held in late April, resulting in conviction of Tom Hand and a sentence to three years in the penitentiary.

District Attorney Key was later involved in an extramarital affair with the wife of Congressman Daniel Sickles, who shot and killed Key when he found this out. Sickles was the first defendant found not guilty in the United States for reasons of temporary insanity. He punished his wife by forgiving her and taking her back. Sickles was later a successful Civil War general.

In 1868, Congressman Charles O'Neill attempted to get the snuff box returned to the nieces of the minister to whom it had been originally presented.[10] He was informed that the recovered items were some loose diamonds and pearls, some melted gold, and a bottle of attar of roses. These were placed in a sealed box and deposited in the United States Treasury, where they then remained.[11] In 1873, the Secretary of the Treasury wrote to the Secretary of the Interior, indicating that these items had been deposited by the Commissioner of Patents on May 10, 1849, and that the space they were occupying was needed.

In 1868, the sword with the jeweled hilt was stolen but was recovered five days later.[12]

[10]See Charles O'Neill to O.H. Browning 2/10/1868, National Archives, Washington, D.C., Record Group 48 entry 138.

[11]See A.M. Stout to O.H. Browning 2/11/1868, National Archives, Washington, D.C., Record Group 48 entry 138; New York Times 3/11/1868 5:2.

[12]See [Kei74].

Chapter 22

The Most Successful Patent Law Firm Ever

Munn & Co., the Scientific American Patent Agency, was partially responsible for the rapid growth of the American patent system. In its percentage-peak years around 1860, one-third of all patents issued by the U.S. Patent Office were prosecuted by Munn & Co.[1] During the year 1865, Munn & Co. filed 3,500 applications in the United States Patent Office, while the whole number of applications filed in the British Patent Office by anyone was 3,000. By 1924, more than 200,000 patents had been issued on applications filed by Munn & Co.,[2] which was more than one-seventh of all the patents ever issued by the United States Patent Office through that date.

The law firm which prosecutes the highest percentage of the issued patents in the present-day (2016) Patent and Trademark Office handles one percent of the patents which issue each year. However, the law firm of Munn & Co., which went out of business about 1960, prosecuted over two percent of all patents ever issued by the United States Patent Office from day one to the present date, even though about two-thirds of all United States patents have been issued since the firm's demise. But we are ahead of our story.

Rufus Porter (1792–1884) was a versatile if eccentric Yankee, who was by turns a portrait-painter, schoolmaster, inventor and editor. He started a small weekly journal called Scientific American, and published his first issue on August 28, 1845. His entire circulation was only a few hundred, and he was bored with the project. He offered it for sale.

[1]See [Mun60].
[2]See [Edi23].

Orson D. Munn

Alfred Ely Beach (1826–1896) was a son of Moses Yale Beach, founder and proprietor of the New York Sun and an inventor of some importance. Beach had attended the Monson Academy, at Monson, Massachusetts, and one of his good friends among his classmates was Orson Desaix Munn (1824–1907). Alfred Beach was working on his father's newspaper and Orson Munn was running a general store at Monson when Beach heard that Porter wanted to sell his journal. The two schoolmates agreed to buy the journal for a few hundred dollars, and they took over in July 1846.

Although Rufus Porter had offered patent advice in the journal when he edited it, Munn and Beach published a paper chiefly devoted to patents and inventions, and they published a weekly list of all patents issued by the Patent Office with claims annexed, and they did so promptly. Never before had information about all patents been available to all the public so quickly. Because of this, they found themselves in the midst of inventors and questions about patents. So they set up the Scientific American Patent Agency to secure patents for their subscribers. Because they were both competent and honest, they succeeded and prospered.

Alfred E. Beach

In 1849, a spare-looking humble man named Allen Benjamin Wilson (1824–1888), a journeyman cabinet-maker from Pittsfield, Massachusetts, came into the offices of Munn & Co. in New York City. The Scientific American Patent Agency offices were in modest quarters which gave Mr. Wilson a sense of security. He carefully untied a handkerchief and took out two models—one for a sewing machine and the other for a rotary steam engine. He said that he was too poor to take out patents on both inventions, and he sought the advice of Munn & Co. about which project to proceed with. Although he obtained a patent on his sewing machine, he entrusted his business affairs to unprincipled men and was cheated out of his invention. However, he set to work again and produced an almost perfect sewing machine, which

Munn & Co. patented for him. This time, in association with Nathaniel Wheeler, he established a prosperous sewing machine company in Watertown, Connecticut, where the once-poor inventor with a handkerchief full of inventions lived in a beautiful mansion in 1858.[3]

Before Munn & Co. came along, most of the patent business in New York City was in the hands of George Garrett Sickles (1797–1887) and Seth Staples (1776–1861). George Sickles was the father of Daniel Edgar Sickles (1819–1914), a lawyer who also practiced patent law. Congressman Daniel Sickles was discussed in Chapter 21.

Circulation of the journal had been less than 300 when Munn & Co. bought it, but reached 10,000 by 1848, 20,000 by 1852, and 30,000 by 1853. In 1849, Mr. Beach personally paid to have a short passenger-carrying pneumatic subway built in New York City, and was charging a fee for the ride until Tammany Hall shut down his subway.

For many years the journal refused much advertising. It limited the amount of advertising that it would run per issue, it would not accept any ad more than 16 lines long, and it would not print a cut in any advertisement.

Allen B. Wilson

Thomas A. Edison said that he used to walk three miles a week as a boy to get his copy of Scientific American.[4] In 1877, he came into the office and placed before the editors his small machine, about which he was willing to make very sparse preliminary remarks. He then turned a crank, and to the astonishment of everyone present, the machine said: "Good morning. How do you do? How do you like the talking box?" This was the first public audience to which the modern phonograph ever addressed itself. A good article on the phonograph was published almost immediately.

Munn & Co. was carefully non-political except to promote the best interests of the Patent Office and the patent system, wherever they might be. They wrote letters with advice to Congressmen. They wrote tutorial letters to Presidents. When they spoke, the government frequently listened. They hired at least one ex-Commissioner to be an attorney in their firm.

[3]See [Edi58].
[4]See [Edi15].

And they hired Robert Washington Fenwick (1832–1896), son of Robert Welsh Fenwick (ca 1804–1845). He had begun his patent training in 1848 at age 16 in the Washington offices of William Parker Elliot, architect and patent agent, who was a former comrade in the 1820s Patent Office of young Fenwick's late father. He also trained in the offices of Zenas C. Robbins, patent solicitor, where he made original drawings. He once testified that he had been associated with the patent business since he was 13 or 14 years old. Robert Washington Fenwick then went to New York City by about 1849 with his letter of recommendation from Mr. Elliot in hand to work for Munn & Co., the Scientific American Patent Agency.[5]

Throughout the 1850s, Mr. Beach traveled to Washington every two weeks to attend personally to applications which the firm had filed in the Patent Office. But a branch office seemed necessary to avoid the time spent on the trip. Munn & Co. soon opened a branch office in Washington, across the street from the Patent Office, to do searches and conduct interviews, as well as to prosecute entire cases coming in through that office. Robert Washington Fenwick was placed in charge of the Washington branch office in 1857. Munn & Co. soon maintained other offices or liaisons with other offices throughout the world.

[5]See [Co.94] and Congressional Information Services HR Committee Hearings 1863–65 (38)HS-T.1-T.6 2/16/1865.

Chapter 23

Commissioner Thomas Ewbank, Historian

In March 1849, after the Patent Office was transferred to the Interior Department, Commissioner Burke was proposing to remove William P. N. Fitzgerald as an examiner for reasons of a personal nature, a feat that he had been prevented from doing when under the State Department. Mr. Fitzgerald had been the assistant examiner under Charles M. Keller, and he had been promoted to principal examiner upon Keller's resignation. Charles Keller wrote a letter to Senator William H. Seward, indicating that Fitzgerald's removal would be a decided injury to the Patent Office.[1]

Charles M. Keller, now a New York City patent attorney, was proposed to Secretary Ewing by Senator Seward, in April 1849[2] for the office of Commissioner, but he requested that his name be withdrawn, because he could not afford to give up his patent law practice, which was far more lucrative. He wrote of the type of man who would, in his opinion, make a good Commissioner. One good choice, in his opinion, would be John Wilson Farrelly of Pennsylvania, former Chairman of the House Committee on Patents.[3] Another

[1]See Charles M. Keller to William H. Seward 3/7/1849, microfilm papers of William H. Seward.

[2]See William H. Seward to Thomas Ewing 4/2/1849, manuscript papers of Thomas Ewing, Library of Congress, Washington, D.C.

[3]See Charles M. Keller to George Perkins Marsh 4/27/1849, microfilm papers of William H. Seward.

man approved and recommended for Commissioner by both Keller and Seward was Thomas Ewbank.[4]

Charles M. Keller

Edmund Burke was notified through the columns of the *National Intelligencer* on May 10, 1849, that Thomas Ewbank had been appointed Commissioner of Patents. While this could not have been a complete surprise, it seems an impersonal way to notify someone that he is out of office. Burke wrote a letter to the Secretary of the Interior asking when he should vacate his office.[5]

Thomas Ewbank (1792–1870) was born at Barnard Castle, Durham, England, and at age 13 he was apprenticed to a local tin and copper smith. When he was 20, he went to London, where he was employed making cans for preserved meats. He saved his money for books and spent seven years pursuing a comprehensive course of scientific study. He was a member of several learned societies when he left England in 1819 to come to New York. He opened a business making lead, tin and copper tubing, which he ran until 1836, then sold out for enough money to allow him to follow his private interest in science for the rest of his life. He wrote and published a widely read book on hydraulics in 1842 and was working on a volume on his explorations in Brazil when appointed Commissioner of Patents. He later finished and published his book on Brazil in 1857. He is widely reported to have written an account entitled "Reminiscences in the Patent Office, and of Things and Scenes in Washington" in 1859, but if he actually wrote it, it is extremely well hidden and has not been discovered by your present guide.[6]

The annual reports of the Patent Office during Mr. Ewbank's tenure reflect the nature of the man who made them. Mr. Ewbank was an author, and when given the opportunity he collected the history of patents and invention in England and in colonial America, published copies of rare pamphlets on the Rumsey-Fitch controversy, and wrote to the governors and senators of the states and territories to ask for copies of their relevant records for publication. The Annual Reports for 1849 and 1850 are filled with historical information

[4]See Charles M. Keller to William H. Seward 4/27/1849, microfilm papers of William H. Seward.

[5]See Edmund Burke to Thomas Ewing 5/10/1849, manuscript papers of Thomas Ewing, Library of Congress, Washington, D.C.

[6]See [Edi70b].

collected by Commissioner Ewbank, much of which would be lost or forever submerged nearly a century and a half later if the collection had not been made when the mass of data in archives and libraries was smaller and fresher.

The Act of March 3, 1837, made the Chief Justice of the District of Columbia a court of appeal from the decisions of the Commissioner of Patents. The occupant of that office was Judge William Cranch (1769–1855), whose uncle, John Adams, appointed him to a junior seat on that bench on his last night in office in 1801. Another of Adams' last-minute appointments, made at the same time, led to the celebrated Supreme Court case of Marbury v. Madison. Judge Cranch became Chief Justice of the District of Columbia in 1805 and held that position until his death at the age of 86. Judge Cranch was paid $100 a year from the Patent Fund, which Ewbank did not considered adequate in 1849, in view of the amount of work involved. In the Annual Report for 1851 was a copy of a letter dated December 15, 1851, from William Cranch to Thomas Ewbank, stating that his great age made it impossible for him to discharge the duties imposed on him by the patent laws. But he never did retire, because there were no government pensions in those days, and although he was 82, a judgeship was for life. The judge's son, William G. Cranch began occasional work as a copying clerk in the Patent Office as early as 1826 and was eventually a patent examiner. In the Annual Report for 1850, Samuel P. Bell, the office machinist, complained that the great hall on the upper floor of the only wing yet built of the Patent Office was occupied by the collections of the Wilkes exploring expedition and the collection of the National Institute. These collections used space needed for about 8,000 of the 12,000 patent models which could not be exhibited for lack of space. The east wing was being built from 1849 to 1852, to provide the necessary space. Not much later, from 1852 to 1856, the west wing was built, and the north wing from 1856 to 1867.

Thomas Ewbank

By the 1851 Annual Report, Commissioner Ewbank complained that the office was so crowded that the mail had to be made up in the halls, where everything was unavoidably exposed to public view. The exhibition of models as contemplated by the 1836 act was not only impossible, but it was also scarcely practicable to protect them from serious injury. He urged that the office space that was being used for exhibition of the collections of the exploring expedition and the National Institute should be cleared and made available for patent models as originally intended. To understand the intense difficulty that this presented for the office, it must be remembered that patents were not printed at this date, and the only way that either examiners or the public had to examine the prior art was to browse through the properly classified models. And, of course, an increase in the clerical force and their salaries was again considered necessary.

Silas H. Hodges

For about five months, from early November 1852 to late March 1853, Silas Henry Hodges (1804–1875) was Commissioner of Patents. Afterwards, he was an examiner-in-chief on the Appeals Board. He faced the usual problems of resignations of examiners for higher pay on the outside, lack of room and facilities for transacting business, and encroachment by the Interior Department on the space and perquisites of the Patent Office.

In the Annual Report for 1852, Commissioner Hodges wrote: "Looking at what has already been accomplished, some find it difficult to conceive that this flood of discoveries and improvements is still to maintain its progress, and even exhibit a swelling tide .... These apprehensions, it is plain, are not to be fulfilled in our day. At no time have there been more decisive indications that every step that is taken in this field ... does but open new scenes to explore and prepares the way for new triumphs."

Interlude for Fiction—A Scene in the Patent Office

From *The Knickerbocker Magazine* of January 1851 comes a short story written by Henry Hiram Riley (1813–1888), with original punctuation. Henry Riley was a lawyer, newspaper editor, prosecuting attorney, Michigan state senator, delegate for Douglas to the 1860 Democratic Convention, maybe a Michigan circuit court judge, and a frequent contributor to *The Knickerbocker Magazine.*[1]

A SCENE IN THE PATENT-OFFICE By an Old Contributor

"This improvement," said my loquacious acquaintance, "will dispense with nine tenths of the wood now used, and will be considered in less than two years the wonder of the age!"

This remark was made to me in the Patent-Office at Washington, by one of those eccentric characters whose life had been made up by piling invention upon invention, all of which, as far as they were modelled and patented, were set up to be gazed at by the loungers who visit that great depository of American genius.

My acquaintance, unfortunately, had spent his whole life for his country, but not one moment for himself. The fruits of his intellect, and the labor of his hands, were very abundant all over the Patent-Office, but at home his wife was out of flour, and himself and children were in rags. The truth is, he never had a moment's time to bestow upon his own immediate needs, for Science had taken him captive, and had driven him for more than thirty years, under whip and spur, through all the mysteries of mechanism. He was a mere originator of complex machines, but never carried anything into practical operation, as that was mere drudgery. Invention was poetry to him, but his mind was satisfied,

[1]See [WF88].

and his stimulus vanished, when he fully became convinced that his labors were successful.

"Nine tenths of the wood," said I; "why, that's nothing! you can't test your invention before the wood and the stove will both be dispensed with. This is a great age, Sir, in mechanics. Twenty, thirty, forty years ago were great ages too, or were then so called. The greatness of a people in any department of science is determined by the advancement of preceding generations, not by the perfection of the age itself; for we do not yet know what the point of perfection in science is, or where or when it may be found. Old Uncle Ben Franklin, great as he was in his day, would, if he should suddenly appear among us, be a mere boy in science."

My acquaintance, however, who had by this time grown warm upon his favorite subject, opened upon me with his argument, in which he attempted to show that science had nearly reached perfection; that first principles were all discovered, and that nearly every application and combination of them that were or ever could be useful to mankind, were already made; in fine, that the intellect of the present generation would in all probability use up all the matériel which nature has provided for it to feed upon; and that in about twenty years, all physical and scientific truths would be rolled up like an old blanket, marked 'demonstrated,' and filed away for the inspection of fools that might follow after.

While listening to his harangue, the walls around me began to expand wider and wider, and the ceiling above raised to an enormous height, while through open doors or passages I saw room after room groaning with thousands of models, until it appeared as though I were in a wilderness of miniature machinery. Very soon a pert little gentleman, with a quick black eye, and a 'pussy' body, arrayed in the queerest costume I ever saw, came bustling up to me, and asked me for my ticket. I involuntarily thrust my hand into the depth of my breeches-pocket, and pulling out a card, delivered it to him. After looking at the card, and then at me, and then at the card again, he burst out into a loud guffaw, that made the old Patent-Office ring. "Why, Sir," said he, "this is no ticket. It is the business card of one John Smith, advertising a patent dog churn, of which he says here he is the real inventor, and it bears date in the year 1850—nearly two hundred years ago! The churn may be found in room marked 'Inventions of the Year 1850," but the man John Smith we haven't got. I don't much think he is around above ground, just at this time,' said the little man, chuckling. "But," said I, "who are you, if I am not John Smith? Were you not appointed by Fillmore Secretary of the Interior, and did I not put a word in his ear favorable to you?" "Fillmore! a Secretary of the Interior!"—exclaimed he; "I appointed by Fillmore! Why, my dear Sir, I was appointed only two years ago—not two hundred!—Chief of the 'Great Central Department,' as the office is now called."

While we were talking, Franklin, Adams, Jefferson, and Fulton walked in and took seats. I knew Uncle Ben the moment I cast my eyes upon him. He was dressed in good old '76 style;—shoe-buckles, short breeches, queue and all; and that same jolly round face and double-chin; that tranquil countenance just touched, without being destroyed, by comedy—were all there. Adams and Jefferson I had before seen, and they were a little more modern in dress, but they both looked care-worn. Fulton sat apart, and eyed the other three as though he had seen them somewhere, but yet could not call them by name.

The rather unexpected arrival of these gentlemen broke up the comments of my bustling interrogator, and one of those pauses occurred which frequently do, upon the appearance of strangers. Uncle Ben asked Jefferson if he would "not like to move up to the fire and warm his feet?" "Fire!" said I, "fire? Why Uncle Ben, there is no fire-place now-a-days. Stoves and hot-air furnaces are all the go. This building is warmed by a great furnace, and two miles of pipe that conducts the heat to every room in it." "Not by a long way!" said my bustling friend—"not by a long way, Mr. John Smith. This trumpery is all piled away among the inventions of the years that were. These things belong to the age of your dog-churn. Why, gentlemen," continued he, "have you never heard of the Great Southern Hot Air Company, chartered in 1960, whose business it is to furnish warm air from the South to persons at the North; price to families three dollars a year; all done by a gigantic underground tunnel, and branches, worked at the other end by an air-pump! Have you never heard of this, gentlemen? Here we get the natural heat of the South, warmed by the sun; none of your stinking coal and wood gases to corrupt and destroy it. And then the principle of reciprocity is kept up; for we send back our cold air in the same way; and so we keep up an equilibrium, for the South are just as strenuous as ever to keep up the equilibrium of the Union. Why gentlemen, those stoves required constant care. As often as every week it was necessary to replenish them with wood or coal. No! no!—those improvements belonged to the dark ages."

"Bless me!" exclaimed Uncle Ben. "Impossible!" repeated Fulton. "And so you don't use the old 'Franklin' stove any more?" said Uncle Ben. "Perhaps," he continued, a quiet smile playing over his face, as if he intended a comical shot, "perhaps you don't use lightning now-a-days either, and my lightning-rods of course belong to the dark ages too!"

"We have the lightning, and use it too, but only one rod, built by the State, near its centre, which is so colossal and powerful that it protects everything around it." And then the little fellow rattled on about the use of lightning;

how it wrote all over the world the English language, until I verily believe that Uncle Ben, Fulton, and all set him down as the most unscrupulous liar that they had ever met with.

"I think,' said Uncle Ben, "that I could convince myself of the truth of your assertions, if I could go to Boston; but as my time is very limited, I cannot."

"Send you there in five minutes by the watch!" answered the little man; "or if that's too soon, in twenty-four hours. It requires powerful lungs to go by balloon—time five minutes—departure every half hour. The magnetic railway train will take you through in four hours, or on the old fashioned railroad in twenty-four." "What!" said Uncle Ben, "is the old stage company entirely broken up?" "Don't know what you mean by stages," said the little man, "but I will look for the word in the big dictionary." "Go by steam-boat," said Fulton. "Steam-boat!" repeated the little man—"steam-boat! too everlasting slow—not over twenty-five miles an hour—well enough for freight, but passengers cannot endure them; they go laboring and splashing along at a snail's pace, and they are enough to wear out any man's patience. Yet the steam-boat was the greatest stride ever made at any one time in the way of locomotion, and was very creditable to Fulton, and the age in which he lived." "That is admitting something," burst out Fulton, who had sat like a statue, watching the little man's volubility. "Men and their works," continued Fulton, "must be judged by the period in which they lived. Each improvement, as it succeeds the last, is aided by its predecessor, and altogether they make out the chain of science." "But," said Uncle Ben, "all this talk don't get me on my way to Boston. That is my birth-place. I was there for the last time in 1763, and you know that according to the provisions of my will, there is more than four millions pounds sterling of my money, which has by this time been disposed of by the State somehow." Uncle Ben was always a shrewd fellow in the way of dollars and cents, and I could see he was very anxious about that money. "Oho! oho!" said the little man; "so you are Ben Franklin, and you are the old gentleman who left that legacy. We've got a portrait of you up stairs, more than two hundred years old, and it does look like you. Glad to see you! You said something in your life-time about immersing yourself in a cask of Madeira wine with a few friends, and coming to the world in a hundred years again. These are your friends, I suppose?" "These gentlemen," replied Uncle Ben, "are John Adams and Thomas Jefferson, signers of the Declaration of Independence." "The other gentleman," continued I, "is Robert Fulton, whom you have spoken of." "Well, I declare!" ejaculated the little man, "this is a meeting! But about that legacy, Uncle Ben, of yours; two millions sterling of it has gone to build the Gutta Percha Magnetic Telegraph line, connecting Boston with London and Paris, two of the largest cities in the Eastern Republic of Europe." "Gutta percha! magnetic telegraph!— Republic of Europe!" repeated all of them. "All built under water, and sustained by buoys," continued the little man, "and it works to a charm—plan up stairs in room 204—and can be seen in a moment; and as I told you before, it writes the English language as fast as my deputy." "Republic of Europe!" exclaimed Jefferson, again. "Yes, Sir," said the little man, "for more than a century. No more thrones; no more rulers by divine right; no more governments sustained by powder and ball; no lords nor nobles; man is man, not merely one of a class of men, but individually man, with rights as perfect and powers as great as any other man. The principles, Jefferson, of your Declaration, which you did not create, but only asserted, have prostrated every arbitrary government on the globe. Even the Jews, since their return to Jerusalem, have organized a republican form of government, and have just elected Mr. Noah President." "Well," thinks I to myself, "that can't be Mordecai M. Noah,[2] anyhow, for politics must have used up his constitution before this." But the little man chattered away, and declared that Europe was divided into two republics, the Eastern and Western; that Constantinople was the capital of the Western; that Africa and Asia were also republican; until the three signers of the Declaration, perfectly wrought up to a phrenzy of joy, rose up from their seats, took off their hats, and swinging them round, gave, "Three cheers for '76, and the old Army of the Revolution!"—and I verily believe Uncle Ben forgot all about that money, and about going to Boston, for he did not allude to it any more in my presence.

"Great changes these!" continued the little man, "from your days. But you must not think, gentlemen, that we have forgotten you or your services, while we have improved in wisdom and strength. Look here, gentlemen," and he motioned us away, and leading on, he conducted us to an observatory on the top of the building. Such a prospect I never before beheld. Away, around, on every side, stretched a mighty city, whose limits the eye could not reach. Towers, temples, spires and masts succeeded towers, temples, spires and masts, until they were lost in the distant haze. Canals traversed every street, and boats of merchandise were loading and unloading their freights. Steam-carriages were puffing along the roads that ran by the canal, some filled with pleasure parties, and some laden with goods. Turning my eye to an elevation, I saw fifty-six gigantic monuments, whose peaks were nearly lost in the sky, ranged in a line, all alike in form and sculpture. "These," said the little man, "were erected to the Signers of the Declaration of Independence;" and taking out his telescope, he handed it to Uncle Ben, who read aloud among

[2]Reference to Mordecai Manuel Noah, (1785–1851), American politician, playwright and journalist. Founded *New York Enquirer*. Was high sheriff of New York City. Attempted to re-establish the Jewish nation.

the inscriptions the names, Franklin, Jefferson, Adams! "But let us know what this city is called? inquired Jefferson." "This, Sir, is called Columbiana; it lies on the west bank of the Mississippi, population five millions, according to the last census." "But what supports it?" "Supports it! The great East India trade. That vessel down there is direct from Canton, by ship canal across the Isthmus. All Europe is secondary to us now. No doubling capes, as was done in your day. Yonder stands the Capitol; and the whole North American continent is annually represented there. The city of San Francisco alone sends forty-four members. There," continued he, pointing his finger, "that balloon rising slowly in the sky has just started for that place, and the passengers will take their dinner there to-morrow."

Jefferson asked the little man "whether the Federalists or Democrats were in power?"—and I saw that Adams waked up when he heard the question. "Don't know of any such division,replied he. "The great measure of the day, upon which the parties are divided, is the purchase of the South American continent at five hundred millions of dollars. I go for it; and before another year the bargain will be consummated. We must have more territory—we haven't got half enough. Extent of territory gives a nation dignity and importance. The old thirteen States of your day, gentlemen, was a mere cabbage-patch, and should have been consolidated into one State. Ten or twenty days' sail ran you plump into a hostile port, and then you had a demand for duty. Besides, conflicting interests always brew up difficulties, and then come treaties, and finally war, and then debt, and at last oppressive taxation. A nation should own all the territory that joins it. The ocean is the only natural boundary for a people." Thinks I, "You have been a politician in your day, and I'll just engage you to correspond with a certain New-York editor,[3] who shall be nameless; you strike off the doctrine boldly!"

Uncle Ben told the little man, after he closed, that a nation might 'get so very ripe as to become a little rotten; and if he had no objection, he would present him with the 'Sayings of Poor Richard.' And suiting the action to the word, he pushed his hand into his breeches pocket, and pulled out an old almanac, printed at Philadelphia, in 1732, and bowing, handed it to him. The little man thanked him, and promised to deposit it in the Museum, as a curious piece of antiquity.

[3]Possible reference to Horace Greeley (1811–1872), who founded the *New York Tribune* and later said "Turn your face to the great West, and there build up a home and fortune." Or possible reference to John Louis O'Sulllivan, of *United States Magazine* and *Democratic Review*, who wrote in 1845 "Our manifest destiny is to overspread the continent."

Getting somewhat anxious for a smoke, I drew forth a cigar and 'locofoco,'[4] rubbed the latter across my boot, which flashed out its light full in Uncle Ben's face. 'That is nice,' exclaimed he; 'rather an improvement on the old string, wheel and tinder plan.' 'Simple, too, isn't it?' said I; 'and yet all the science of your day didn't detect it.' Just then I gave a puff, which made Uncle Ben sneeze; and he broke out in a tirade against tobacco, that would read well. But I told him there was no use; men had smoked and chewed the weed—would smoke and chew it, economy or no economy, health or no health, filth or no filth; and that in all probability the last remnant of the great American Republic, for succeeding nations to gaze at, would be a plug of tobacco; for I sincerely believed that tobacco would outlive the government itself.

The little man proposed returning into the Patent-Office, and exhibiting to us in detail the models of the art there deposited. But I cannot weary the reader with what I saw there. The fruits of every year, since the organization of the department, were divided into rooms, and indicated on the door by an inscription. There were thousands of improvements in every branch of science, many of which were so simple, that I thought myself a fool that I did not discover them long ago. Principles were applied, the very operation of which I now recollected to have often seen, yet without a thought of their practical utility. I came to the conclusion that accident was the parent of more that I saw than design; 'for how,' reasoned I, 'is it possible that these pieces of machinery could otherwise have escaped the great men who have lived and died in ignorance of them?'

By this time we were quite fatigued, and Uncle Ben complained a little of the 'stone,' which he said he was subject to. The little man gave him some 'Elixir of Life,' as he called it, being, as he said, 'an extract of the nutritious portion of meats and vegetables, purged from their grossness as found in their natural state;' and while we were sipping it, he launched forth upon its great benefit to mankind; the money saved that used to be expended in cookery and transportation—millions upon millions; the great economy in time, formerly squandered in eating, etc., etc.; and he wound up his eulogy by presenting each of us with a bottle, which I carefully put away in my pocket.

Adams then rose up, and said he must leave, and Jefferson, Uncle Ben, and Fulton followed. 'But,' said Jefferson, 'I have a word to say on my departure. There is one thing, of more value than all I have seen, for it is the father of all: you should reverence it next to the Creator of all the Universe. Overlook it not in prosperity, nor despair of it in adversity. It is The Union. Better perish

[4]'Locofoco' by this time meant a friction match. Originally, it was a term applied by John Marck, an inventor (1834), to his self-lighting cigar. The term is loosely interpreted to mean self-firing.

with the Union, than survive its ruin!' And in a moment Uncle Ben, Fulton, Adams, Jefferson, the little man, the apartments, wheels and machinery began to rock, and heave, and fade, and finally dissolve; and suddenly I awoke! It was a dream!—and there I sat, my tormentor affirming that his stove was perfection, that it would save three-fourths of the wood, etc., etc., until, out of patience, I pronounced him a blockhead, gave him a kick, put on my hat, and departed.

Constantine, (Mich.) H.H.R.

Chapter 24

Judge Charles Mason's Patent Office

Commissioner Hodges was succeeded by Charles Mason (1804–1882), who was generally believed to be the most effective Commissioner of Patents to that date and perhaps the most effective Commissioner in the nineteenth century. Mr. Mason graduated first in the United States Military Academy class of 1829. (Number two in that class was Robert E. Lee.) He then served three years as a professor at the Military Academy. He left the Army to study law and eventually settled in Burlington, Iowa. He was appointed Chief Justice of the Territory of Iowa in 1838 and held that position until 1847. Charles Mason was appointed Commissioner of Patents on May 16, 1853. While Judge Mason was a brilliant and hard-working man, his success in office is due as much to the people he surrounded himself with as to his own abilities.

William Chauncy Langdon (1831–1895) was a very bright young man who graduated from Transylvania University at age 16 and almost immediately became an adjunct professor in chemistry and astronomy at Shelby College. He was appointed an assistant examiner in the Patent Office before he was twenty.[1] While still an examiner, he was one of the founders of the Young Men's Christian Association in America, and when he later resigned from the Patent Office, he practiced briefly as a patent agent before becoming a Protestant Episcopal clergyman in 1858.

[1]See Joseph Henry to Charles Mason 3/2/1853, manuscript papers of William Chauncy Langdon, Library of Congress, Washington, D.C.

William C. Langdon

Mr. Langdon was actually appointed to the Patent Office in May 1851 in Commissioner Ewbank's tenure, to work under examiner Homer Lane.[2] Once, in August 1852, when Langdon had just returned to the office from being ill at home, Commissioner Ewbank came to him and, in a good-natured way, berated him for taking off from work just in time to have a disagreeable person show up to interview the Commissioner because the examiner was not present.[3]

On June 30, 1853, Commissioner Mason promoted William C. Langdon to examiner third class. In September 1853, William Langdon was offered a bribe of $20 in a letter from Charles T. Porter of New York. When this letter was reported to the Commissioner, Langdon noted[4] that Judge Mason had written such a letter that "that man won't try to bribe an examiner again." By the end of September, he was appointed an acting principal examiner,[5] and by 1855 he had been appointed a full principal examiner.

In March 1855, Mr. Langdon was lobbying on behalf of the Patent Office and the examiners to secure payment of a total of $14,000 back pay for examiners. This amount was to be funded by the device of having an appropriation voted to restore to the Patent Fund money which had been used from that fund over the years for purchase of seeds for distribution by the Patent Office and then making the money available for back pay and raises.[6] Although the raises were approved without difficulty, a clerical error in the conference committee resulted in raising the pay for the chaplain of the penitentiary rather than providing back pay for examiners.

[2]See Homer Lane to William Chauncy Langdon 5/22/1851, manuscript papers of William Chauncy Langdon, Library of Congress, Washington, D.C.

[3]See Langdon diary 8/13/1852, manuscript papers of William Chauncy Langdon, Library of Congress, Washington, D.C.

[4]See Langdon diary 9/14/1853, manuscript papers of William Chauncy Langdon, Library of Congress, Washington, D.C.

[5]Langdon to his mother 9/21/1853, manuscript papers of William Chauncy Langdon, Library of Congress, Washington, D.C.

[6]See Langdon diary 3/3-6/1855, manuscript papers of William Chauncy Langdon, Library of Congress, Washington, D.C.

On March 21, 1855, William De Hartburn Washington (1834–1870), a Patent Office draftsman from Virginia and New York, drew an unauthorized picture in colored crayons on the whitewashed wall of one of the rooms in the basement of the Patent Office. It was a copy of Leutze's Washington Crossing the Delaware. The Patent Office messenger, George R. Adams, saw it and was getting ready to whitewash over it when Langdon praised it. Soon Commissioner Mason saw it and brought the President and the Secretary of the Interior in to see it. The next day, President Pierce brought Mrs. Pierce and another lady in to see the picture. The artist became a painter of portraits and historical subjects. The picture was still in place in October 1856.[7]

Clara Barton

Titian R. Peale continued in the Patent Office under Judge Mason. He had been passed over for principal examiner on every occasion since his initial appointment in 1848. One of Commissioner Mason's promotions in his first few months in office was a promotion of Titian R. Peale to principal examiner.[8]

Charles Mason

Another of Judge Mason's appointments that has to be regarded as a success is that of Clara Barton. Yes, that Clara Barton. To begin with, Clara Barton (1821–1912) had resigned her job teaching school in Massachusetts and had come to Washington to seek work. Commissioner Mason intended to hire Miss Barton to go out to Burlington, Iowa, to be a governess to his daughter, there then being no schools in Burlington.[9] However, she had the capacity for a much more difficult job, and Judge Mason hired her as his confidential clerk at a salary of $1,400. She thus believed that she was the first woman ever to be hired to a regular position in the U.S. government with work and wages equal to that of a man.[10]

[7] See *U.S. Magazine* 10/1856 pp. 289–298 and [GW57].

[8] See [Edi53].

[9] See Charles Mason Remey note.

[10] See [Bar22].

There had been leaks of confidential information from the Patent Office, with some dishonest clerks selling secrets, resulting in injury to the department and to owners of patents. When she was confidential clerk, these leaks stopped. Miss Barton did rouse jealousy among the male employees in the Patent Office, who were impolite to her on any possible occasion. During the temporary absence of Judge Mason from the Patent Office in 1855, she was removed from office by the Secretary of the Interior, but Mason gave her a temporary appointment upon his return. Sometimes, she worked in the office, and sometimes, when there was too much political criticism of her presence in the office, she worked at home. Much of the work which she did at home was copying work, copying letters and patents into large bound volumes.

In the Annual Report for 1854, Judge Mason reported that his office had reduced the number of pending applications in 1854 from 823 to 89 and had doubled the number of patents issued from the previous year. In 1854, the office issued a record 1,902 patents. The previous high in 1849 was 1,070 patents in one year. While the Annual Report for 1853 had illustrated the issued patents with woodcuts, the report for 1854 illustrated each patent, for the first time, with copperplate engravings, adding greatly to the clarity of the illustrations.

Judge Mason noted that the Patent Office had twelve principal examiners in 1854 and issued more than 2,000 patents during the year. There had been a fiction that the Commissioner knew everything that was happening in the Patent Office, but Judge Mason wrote that it was now impossible for the Commissioner to exercise personal supervision over the numberless cases being presented for official action. Whenever one of the principal examiners reported in favor of granting a patent, it was granted without further question.

The Commissioner was concerned that the patent rights of a citizen should depend upon the mere discretion of an executive officer. This seemed to him to be more of an Asiatic system than an Anglo-Saxon system. There should be a right of review of Patent Office decisions by a court—a fair trial by a regular judicial court. It was true that there was a right of review by the judge of the circuit court, but this was not a new hearing. The judge heard the case on the record before the Patent Office and was paid out of the Patent Fund. Judge Mason considered such a judge to be part of the Patent Office for the occasion.

In the Annual Report for 1856, the Commissioner pointed out that the English Patent Office printed copies of all patents, including their drawings, as they issued and sold copies at prices which merely defrayed the expenses. He recommended that the same be done by the U.S. Patent Office. This would enable copies to be furnished to individuals at one-tenth the cost of copying them out by hand, and it would make it possible for the Patent Office to have more than one copy for the use of the examiners. The handmade copies in use in the Patent Office were used so much that it was frequently necessary to copy them as the originals wore out. He wanted a copy of each patent deposited with the clerk of each federal district court to enable inventors to consult them without having to make a trip to Washington for the purpose.

We know more about the thoughts and motives of Charles Mason than of any other nineteenth century Commissioner because he kept a diary from June 5, 1855, until February 16, 1882, nine days before his death.[11] In the first entry, made halfway through his tenure as Commissioner, he noted: "This day a cripple called on me with books for sale. I purchased this with a view of keeping a diary which in all my life thus far I have never done." We owe the unnamed "cripple" a debt of gratitude for his gift to Patent Office history. On the same day, Mason attended a meeting of the "Establishment" of the Smithsonian Institution, but only Professor Henry was present. The Commissioner was ex officio a member of the Smithsonian Institution.

The Commissioner heard appeals from rejections by his examiners, and he frequently reversed their rejections. He noted on June 15, 1855, that he would probably order the issuance of as many patents that year as would any one of the examiners. But by the end of the month, he was planning to resign because of political differences with someone. He sent in his resignation on July 4, 1855. By September 30, he was still trying to have the Secretary of the Interior appoint his successor.

But his resignation did not have the anticipated effect. Judge Mason had suffered from profound melancholia since the death of two daughters four years earlier. The only way he knew to overcome it was to lose himself in work. About September 1855, the Secretary had made some encroachments on the Patent Office to provide space for the Indian Bureau and had been personally attacked by Scientific American. The Secretary had also been dismissing female clerks from the Patent Office during Judge Mason's absence. By November 3, 1855, Judge Mason was back in the office of Commissioner, having officially never left it. On November 7, 1855, the Commissioner had dinner with President and Mrs. Pierce and discussed the Patent Office.

On November 30, the Commissioner found it necessary to dismiss a watchman of the Patent Office because he was a member of the Know Nothing Party.

On December 9, Judge Mason learned from an Iowa member of Congress that his name was being proposed by farmers and mechanics across the country for President of the United States. He was astonished. However, few members of the federal government were as well known to the public as he was. Every year, the Annual Reports were published and widely distributed, both

[11]See Charles Mason diaries, manuscript papers of Charles Mason, Library of Congress, Washington, D.C.

to mechanics for the patent information and to farmers for the agricultural statistics. The agricultural sections of the Annual Reports were widely quoted in newspapers in every farming community. Every year, the Patent Office sent out seed samples for tens of thousands of farmers to experiment with. The name of Charles Mason was firmly and favorably associated with each of these transactions.

There was a spirited and bitter contest for Speaker of the House, and when it was finally resolved by compromise some clerks loyal to the Democratic party had to be removed from the House payroll to make room for those entitled to offices by the compromise. On February 6, 1856, President Pierce required Judge Mason to remove some Patent Office non-Democratic-Party clerks to make room for those displaced in the House by the compromise. Mr. Mason opined that retaliation in politics cannot be complained of by those responsible for it.

Two days later, General Lewis Cass fell down the steps on the east front of the Patent Office, seriously injuring himself. At the time, there was doubt if he would survive. General Cass was then a Senator, but he had been running for President during the election which took place on the night of the great Patent Office jewel robbery in 1848. He did recover from his fall and lived a further 10 years.

In late April 1856, when William Chauncy Langdon was about to resign his position as principal examiner, President Pierce had been urging the appointment as replacement of various individuals that he thought might help him to be re-elected. First, he recommended Samuel McKain, a former clerk in the House of Representatives, but Judge Mason was unwilling to appoint him to a position higher than assistant examiner. Then the President began urging the appointment of a Mr. Street of Virginia, who had been a partisan of President Pierce. On May 6, Mason appointed Thomas H. Dodge of New Hampshire, former assistant to Mr. Langdon, to fill the vacancy, again at the President's recommendation, and to Langdon's great delight. Several years later, Mr. Dodge was to be one of the first three members of the Board of Appeals.

On May 14, Judge Mason called on James Buchanan and pronounced him a splendid man whom he should be glad to see as President. However, he had misgivings concerning some of the men he had surrounded himself with, including Daniel E. Sickles, the congressman and occasional patent attorney, whom he regarded as a political fortune-hunter.

The Commissioner noted with regret the death of William E. Aisquith (ca. 1808–1856), who had been two years ahead of him at the U.S. Military Academy. Mr. Aisquith, while in the Army, had become such an alcoholic that he was cashiered and obliged to leave the Army. During the Mexican War he had served as a lieutenant. Two years earlier he had asked for a position as watchman in the Patent Office, and the Secretary of War had recommended him. Judge Mason hired him as watchman. Mason thought that Aisquith had still indulged in alcohol on occasion but not so much that he had felt impelled to discharge him. His wife had died years earlier, and he left three orphaned daughters.

As an example of the sorts of appeals that were taken to the Commissioner, on May 12 he wrote that he had been puzzling over a philosophical problem relating to a revolving disk which seemed to depart from the ordinary laws of gravity. He could not quite understand what was happening. On July 7, he finally reached a solution to the problem. This was probably issued as U.S. Patent 15,353 on July 15, 1856, for a clothes washing machine in which a revolving disk with pegs on its lower side was placed over a load of clothes in a washing machine to assist in washing the clothes.

On September 9, he noted that he had held a conversation with Secretary of the Interior Robert McClelland in which he decided that the Secretary regarded himself as authorized to control the Commissioner in matters left by law to the Commissioner's discretion, particularly the right to appoint temporary clerks and the right to obtain sugar cane cuttings for distribution.[12] The Secretary also said that no appointment, however trivial, and no project involving the expenditure of money, however small, should be entertained without first consulting the Secretary. By blocking the Commissioner's action to obtain sugar cane cuttings, the Secretary had prevented the Commissioner from complying with the directions of Congress.

Robert McClelland

[12] See Clara Barton to Stephen Barton 9/28/1856 in [Bar22].

New model room of U.S. Patent Office second floor (old style), east wing (1854)

Judge Mason indicated that he was feeling very uncomfortable in his position and did not think he would continue in it for long. On September 15, he sent his resignation in to the President. By September 27, the President called him in and asked him to remain until the Secretary returned to Washington, at which time he was confident that all could be arranged satisfactorily.

In his diary for December 16, 1856, Judge Mason noted: "I am almost overwhelmed by applications for employment. What an alms house is this City of Washington!" On February 16, 1857, Mr. Mason wrote that he had just been offered $6,000 a year by one man to attend his patent business if he left the Patent Office. On July 28, 1857, Judge Mason sent his resignation to the President, to take effect on August 5. He began the practice of patent law. By January 12, 1860, he was working for Munn & Co.

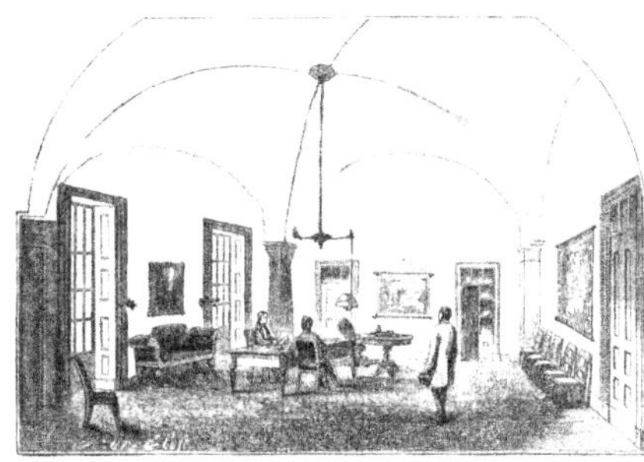

Commissioner Mason's Room in 1856 Patent Office

Chapter 25

Antebellum

Elmer E. Ellsworth (1837–1861), who was no kin to ex-Commissioner Ellsworth, became a clerk in the Chicago office of Arthur F. Devereux, patent solicitor, about 1855, at age eighteen. They soon became partners in the business. The firm was becoming prosperous by the time he was twenty-one. However, after they had entrusted their funds to an agent who has not been named, the agent absconded with the funds, robbing them of the accumulation of three years of work. Mr. Ellsworth then entered the law office of Abraham Lincoln at Springfield, spending part of his time studying law for admission to the bar and part of his time promoting a scheme to reorganize the Illinois militia.[1]

Abraham Lincoln was no stranger to the patent law. In late 1848, while Lincoln was coming up the Detroit River on the steamboat Globe, the boat ran aground on the shoals off Fighting Island. When he returned to Illinois, he began work on a model of a new boat with inflatable bellows on each side of the boat just below the waterline. When this boat ran aground on shoals, the bellows were to be inflated, buoying the boat over the shoals. Abraham Lincoln had been elected to one term in Congress, from March 1847 to March 1849. When he returned to Washington for the second session, he brought his

Abraham Lincoln

[1]See [Ing18] and [Ran60].

model with him, went to the offices of Zenas C. Robbins, patent solicitor, and applied for a patent. One of the fond memories of Robert Todd Lincoln, his son, was the time in 1848 when his father took him to the Model Room of the Patent Office to look at the displays. Abe Lincoln's patent was issued on May 22, 1849, as U.S. Patent No. 6,469, for "Buoying Vessels over Shoals." He thus became the only President who was a patentee. And as noted earlier, at the time when Mr. Lincoln went to the office of Zenas C. Robbins to secure a patent, Mr. Robbins' apprentice draftsman, who may well have worked on the drawings for Mr. Lincoln's application, was our young friend Robert Washington Fenwick.

Early in his career, Lincoln won the unreported case of Parker v. Hoyt for the defendant, proving to a jury that his client's waterwheel did not infringe the plaintiff's patent. His second largest professional fee came from his participation in the Reaper Case, McCormick v. Manny, in association with George Harding, prominent patent attorney of Philadelphia, and another aggressive Pennsylvania lawyer, Edward M. Stanton, representing the defendant Manny. Although Mr. Lincoln was well-prepared and well-paid for his intended argument, Harding and Stanton thought him too ungainly and unpresentable to be allowed to participate in the arguments, and he was sent home unheard. Nevertheless, the defendant won. Lincoln argued his last patent case, Dawson v. Ennis, after his nomination for President, but before his election. Although Lincoln won the election, his client lost his case.[2]

Joseph Holt

Lincoln held no hard feelings toward Stanton and Harding. Upon his election as President, he offered the job of Commissioner of Patents to Harding, who refused it. He later offered the position of Secretary of War to Edward Stanton, who accepted it.

The Commissioner of Patents appointed to replace Judge Mason was Joseph Holt (1807–1894) of Kentucky. Mr. Holt began the practice of law in 1828 in Louisville and served as commonwealth's attorney there from 1833 to 1835. He then moved to Mississippi and practiced law until 1842. When he won a large judgment against the city of Vicksburg on behalf of the heirs of the founder of the city, he became wealthy and gave up the private practice of law. He lived a quiet life back in Kentucky, engaging in much foreign travel, until he moved to Washington in

[2]See [Gol38], [Smi59], [Fos28], and [Ros28].

the spring of 1857. When he was first offered the office of Commissioner, he turned it down, but soon thereafter accepted it. According to the new Secretary of the Interior and one-time fellow Mississippi lawyer Jacob Thompson, Mr. Thompson had obtained the position for Mr. Holt because Mr. Holt was a briefless lawyer with nothing else to do. This and other more unfriendly remarks were made years after a Civil War during which Mr. Holt had been Judge Advocate General of the Union Army and Mr. Thompson had been Governor of Mississippi. But again, we are well ahead of our story.

Joseph Holt was appointed Commissioner of Patents on September 10, 1857. By December, Mr. Holt had established a Board of Appeals to hear appeals from adverse decisions of an examiner. The initial Board of Appeals consisted of Thomas H. Dodge, DeWitt C. Lawrence and A. B. Little. This was not the official Board of Appeals later authorized by Congress in 1861, but it served as an unofficial advisory board to the Commissioner. According to reports in Scientific American, Mr. Holt had been attempting to persuade examiners to be more liberal in granting patents. Applicants had the right to appeal to the Commissioner from adverse decisions of an examiner, but Commissioner Holt was being overwhelmed with appeals from examiners that he could not convince to be more liberal in their decisions. Thus he appointed a Board of Appeals to take the pressure of work off his shoulders.[3] Finally they reported that the Commissioner felt that certain old examiners were attempting to subvert his policies of liberality, almost to the point of insubordination, so he dismissed them. Scientific American likened this to Dogberry's maxim, that when two people undertake to ride the same horse, one, of necessity, must ride behind the other.[4] Commissioner Holt was determined that he should ride in the front, and, as it were, determine where the Patent Office pony was to go. By 1858, Commissioner Holt was complaining that withdrawal of the three examiners from their regular duties had reduced the number of principal examiners in the examining force

1859 Office Seal

[3]See [Edi57] and [otCoP58].
[4]See [Edi60b].

by 25 percent. Mr. Holt wanted more principal examiners to be authorized by Congress.[5]

Early in the administration of Commissioner Holt, Oscar J. E. Stuart, of Holmesville, Mississippi, requested information on how to get a patent on an invention made by Ned, a slave belonging to Mr. Stuart's late wife's estate. The invention was an improved cotton scraper plow for plowing cotton fields. Jacob Thompson presented the matter to Attorney General Jeremiah S. Black for an opinion, but he refused to give an advisory opinion, indicating that he would pass upon an actual pending patent application. When Mr. Stuart recited all of the relevant facts and filed an application in his own name, the Attorney General ruled that Mr. Stuart could not claim to be the inventor and so could not make the application in his own name.[6]

Benjamin T. Montgomery

Ned, being a slave, was not considered a citizen of any country, could not make the oath of citizenship required by the law and so could not make the application in his name. A century or more later, some people have considered this to be another instance of the federal government depriving slaves of rights, but it could also be interpreted as a federal government which deprived slave owners of at least one benefit of owning slaves. In the Annual Report for 1857, Mr. Holt indicated that this was only one of several applications that had been received within the past year for inventions made by slaves, a situation which he believed had never arisen before. Oscar Stuart was later a colonel in the Confederate Army. Nothing further is known of Ned.

There was a similar incident somewhat later, less well documented, in which Benjamin T. Montgomery (1819–1877), a slave on the plantation of Jefferson Davis' brother Joseph Davis, invented a propeller for a river steamboat, operating on the canoe-paddling principle. It is said that both Davis brothers tried to get a patent for Benjamin Montgomery, but they were prevented from doing so by the Attorney General's decision in Ned's case. A few years later, when Jefferson Davis was President of the Confederate States of America, he signed

[5]See [otCoP58].

[6]See Oscar J. E. Stuart to Jacob Thompson 8/25/1857 and 12/18/1857 both National Archives Record Group 48 entry 1 box 1, [Yan84], and [Boy60].

legislation into law allowing slaves to get patents for their inventions. Apparently, Benjamin Montgomery later filed his application for a U.S. patent as a freed man on June 28, 1864, but did not receive a patent. Perhaps this is because of a strong similarity between his paddling propeller and the steamboat that John Fitch demonstrated to the Constitutional Convention.[7]

In March 1859, Commissioner Holt was appointed Postmaster General to replace A.V. Brown, who had recently died. Later in the administration of President Buchanan, he was made Secretary of War. He was the only member of Buchanan's final cabinet who did not side with the Confederacy.

William Darius Bishop

On May 7, 1859, William Darius Bishop (1827–1904) of Connecticut was appointed as the new Commissioner of Patents. Mr. Bishop had been president of the Naugatuck Railroad in 1855, and had served one term as a Congressman from 1857 to 1859, and when his term in Congress expired, he was appointed a 31-year-old Commissioner of Patents. Very early in 1860, Commissioner Bishop arranged for photographic copies to be made of certain patent drawings and used these copies to supply copies to the public and also to provide extra copies for examiners to use in examining applications.[8] This practice was not continued for long. In Mr. Bishop's only Annual Report as Commissioner, that for 1859, he told Congress on January 26, 1860, that he could not comply with their recently imposed limit of 800 pages for an annual report. He said that he was submitting one of 1,200 pages because a shorter report would be useless. Possibly because he left office very shortly after making this annual report, there were no repercussions. Why Mr. Bishop left office so suddenly seems not to have been reported. On January 28, 1860, Scientific American was praising his work and wishing him continued success and by February 4 was announcing the expected appointment of Samuel C. Ingham to be the new Commissioner.

However, Mr. Ingham refused the office, and the new Commissioner was Philip Francis Thomas (1810–1890), ex-Governor of Maryland, who was ap-

[7]See [Jam89], [Her81], [Bak17], *New York World* 9/27/1890 pp. 1–2, *Cincinnati Daily Gazette* 12/25/1863 pp. 3:8, and Commissioner W.E. Simonds to Mrs. Jefferson Davis 5/14/1892 National Archives Suitland Accession no. 55-A-584 Record Group 241 Box 19.

[8]See [Edi60c].

pointed on February 15, 1860. Mr. Thomas resigned from the office of Commissioner on December 10, 1860, to become Secretary of the Treasury in President Buchanan's cabinet. Mr. Thomas was soon a Southern sympathizer during the Civil War.

Philip Thomas

In the summer of 1859, John Mercer Brooke (1826–1906), a lieutenant in the U.S. Navy, was shipwrecked in Yokohama, Japan, where he established good relations with the Japanese government. When Japan was to send a diplomatic mission to Washington the following year to exchange ratifications of the Harris treaty of 1858, Lieutenant Brooke and ten of his seamen were invited to accompany the mission and train Japanese mariners in ocean navigation. The ambassadors, Shimmi Masaoki, Lord of Buzen (1822–1869), and Muragaki Norimasa, Lord of Awaji (1813–1880), left Japan in January 1860 on the American warship Powhatan. A Dutch-built Japanese ship, Kanrin Maru, sailed with the American ship, with Lieutenant Brooke and his men aboard to help train the Japanese crew.[9] When the ambassadors arrived in Washington, they had a full schedule of visits. On May 21, 1860, the ambassadors were given a guided tour of the Patent Office by Commissioner Thomas. The model cases were unlocked, and Commissioner Thomas handed them for closer inspection any models which seemed to interest them. The Japanese seemed to understand everything that was shown to them. There was a great crowd of people following them everywhere they went, observing the strange sight of Oriental men in silk brocade robes, knee britches, and dark stockings, who wore curious hats resembling small houses strapped to the top of their heads. In fact, the ambassadors were so besieged by a crowd that they could hardly move, so they cut short their visit to the Patent Office. However, some of the Japanese made several return visits to the Patent Office to examine models and drawings. Others were frequent visitors to Matthew Brady's gallery, learning the daguerreotype business.[10] One of the extensive crowd noted the visit in his diary.[11] On May 29, 1860, Charles Mason was invited by Professor Joseph Henry to meet the Japanese ambassadors at the Smithsonian. Charles Mason was not the only one to go out of his way to meet the ambassadors— whole

[9]See [Ita83].
[10]See [Edi60d].
[11]See [FCM89].

families travelled 500 miles by railroad to see the fabled visitors from the Far East. The Japanese were unfamiliar with American customs, and Muragaki noted in his diary that another member of his party drank the water in his finger bowl at a state dinner. At the Smithsonian they saw wigs worn by several former Presidents and thought they had seen a disgusting display of scalps removed from the heads of dead men.

Left: Shimmi Masaoki
Right: Muragaki Norimasa

Upon the transfer of Commissioner Thomas to the Treasury Department, Chief Clerk Samuel T. Shugert became acting Commissioner for a period of months. He had acted in this office before, during Mr. Mason's attempted resignation in 1855. Mr. Shugert had been a clerk in the Patent Office since about May 1845,[12] and he was Chief Clerk under Judge Mason. He remained Chief Clerk until removed from the office about June 10, 1861.[13] He made the Annual Report for 1860.

Rufus Randolph Rhodes (1818–1870), a lawyer of Louisiana, was appointed assistant examiner in the Patent Office on June 20, 1857.[14] He was soon a principal examiner and a member of the Board of Appeals. On November 24, 1860, Judge Mason received a letter from W. W. White of Burlington, Iowa, suggesting ways that war could be avoided. Mr. Mason showed the letter to several Southern friends, who said if there were 100 men like Mr. White in the North, they felt they would have hope of avoiding war. When he showed it to Rufus Rhodes, Rhodes said that if he thought there were two men like Mr. White in the North, he would have hope. But hope soon disappeared, and Mr. Rhodes resigned from the office and headed South.

[12]See Samuel T. Shugert to James Buchanan 5/12/1845, manuscript papers of Edmund Burke, Library of Congress, Washington, D.C.

[13]See Mason diary, manuscript papers of Charles Mason, Library of Congress, Washington, D.C..

[14]See Charles Mason to Rufus Randolph Rhodes 6/20/1857 registered vol 4 National Archives Record Group 48 entry 256 and [Edi61f].

Chapter 26

The Union Patent Office

Secretary of War (and ex-Commissioner of Patents) Joseph Holt tried to prevent the need to withdraw from Fort Sumter after the secession of South Carolina on December 20, 1860, and the firing on a ship landing supplies at the fort on January 9, 1861. The fort held for the balance of the Buchanan administration and did not fall until after Abraham Lincoln's inauguration.

The Patent Office was picking up speed. When the Patent Office burned in 1836, 10,000 patents were destroyed, the total issue of 46 years. By 1859, the Patent Office was issuing 10,000 patents in about 46 months.

John Smith showed up again in Washington. He had invented a steam plow this time, and when he came to Washington to promote it, some of the residents at his boarding house told him that to succeed in securing his patent he must make himself prominent in Washington society. Although he was a stranger, his appearance at the President's levee drew considerable attention because of the peculiarities of his costume, which consisted of a brilliantly colored military uniform, with a silk scarf thrown over his shoulders.[1]

Charles Mason, who had joined Munn & Co. soon after leaving the Patent Office, left that firm on good terms on September 25, 1860, and came to Washington to practice law. He wanted to practice general law, but the path of least resistance was to take the patent cases which beat a path to his door. In October, Robert Washington Fenwick proposed that Judge Mason form a partnership with him. Mason was not ready for that but did rent office space with Fenwick. As the world changed around him, Mr. Mason noted it in his diary. He went to a sermon by Dr. Revere Gurley, a prominent Washington minister and ancestor of a late patent attorney of the same name in your

[1] See [Edi60a].

guide's former law firm. He and his daughter went to a performance by the Christie Minstrels (the old ones). He went for a walk across the long bridge into Virginia, thinking about possible solutions for an application that he was at that moment prosecuting. One of his clients was so upset by South Carolina's secession that he could not attend to his business. On December 10, 1860, he noted a violent debate in the Senate which seemed to make reconciliation impossible. The streets of the city were deep mud. The Patent Office, which had always paid its expenses and had a small surplus, was beginning to show a short-term loss. And on January 22, 1861, Judge Mason noted in his diary that senators and representatives from three Southern states had walked out that day and headed home.

Elmer Ellsworth

There was much Southern dissatisfaction with the election of Abraham Lincoln, and it was predicted that assassination attempts would be made before his inauguration. Among the guards who therefore accompanied Abraham Lincoln to Washington was his law clerk, newly admitted to the bar, Elmer E. Ellsworth, the former patent solicitor. Soon after the surrender of Fort Sumter on April 14, 1861, Elmer Ellsworth went to New York to raise a regiment for Union service. He recruited among the firemen of the city, and his 1,100 troops were called the Fire Zouaves. The Fire Zouaves were mustered into service in front of the Capitol in the presence of President Lincoln, and they were soon sent to occupy nearby Alexandria, Virginia. When Colonel Ellsworth quickly learned that an Alexandria innkeeper, James W. Jackson (1824–1861), had raised a Confederate flag from the rooftop of the Marshall House inn, he stormed the inn in the early morning of May 24, 1861, and tore down the flag. Mr. Jackson shot and killed Colonel Ellsworth as he was coming down from the roof and was immediately killed himself. Thus, Colonel Elmer Ellsworth became the first Union casualty of the Civil War, if one does not count the four Massachusetts soldiers killed five days earlier by a Baltimore mob. Colonel Ellsworth was proclaimed a martyr. Clara Barton, still a Patent Office copyist, wrote in a letter that she wondered if Ellsworth had not sold himself at his highest price for his country's good—if the inspiration of his death were not worth more to his cause than the life of any man could be. A regiment was immediately formed in New York City named Ellsworth's Avengers. Scores of families named their babies Elmer Ellsworth, and the name has been carried down to the present day in some of those families. A recent patent attorney with the name of Elmer Ellsworth Goshorn was known to the author.[2]

David Holloway

On March 2, 1861, the term of a patent was changed from 14 years, with a possible extension for seven additional years, to a fixed term of 17 years. On March 28, 1861, David P. Holloway (1809–1883) was appointed Commissioner of Patents. He had been a congressman from Indiana (1855 - 1857) who had tried to establish a department of agriculture. Mr. Holloway was a remarkable man, in that he ran the Patent Office to the apparent complete satisfaction of Munn & Co., which was difficult to do, while apparently being extremely unethical by present-day standards. When Mr. Holloway arrived in Washington, rental houses were very difficult to find because of the war. Daniel Jay Browne (1804-ca 1867) had been the head of the agricultural division of the Patent Office from June 9, 1853, through 1859. Judge Mason had noted in his diary on May 6, 1856, that Browne was in some respects a perfect clerk but in others was perfectly incompetent. Although he lived in Mary Surratt's boarding house in 1855, by 1861 Browne had a handsome house in Washington, which Mr. Holloway wished to rent. A deal was struck, wherein the Patent Office would hire Mr. Browne at a salary of $3,000 per year to travel around Europe to look for information that would be important to farmers and manufacturers in this country. While his Washington house was empty, he would rent it to Mr. Holloway for a quite reasonable $500 per year. The only report made by Mr. Browne during the entire period of employment in Europe was an article on flax, published in the agricultural section of the Annual Report of 1861, and subsequently found to have been copied from an old English book. Mr. Holloway also employed one

[2]See [Ran60] and [Ing18].

of his own sons at $1,600 per year as the receiver of models from applicants for patents and another son as the chief messenger of the Patent Office.[3]

First Rhode Island Infantry Regiment Quartered in Patent Office

On April 28, 1861, Judge Mason noted that he had seen a battalion being paraded in front of the Patent Office and that the soldiers were quartered in the Patent Office. In April and May 1861, the soldiers of the First Rhode Island Regiment were quartered in the model rooms of the Patent Office, together with their laundress and three ladies who utterly refused to be left at home. They eventually moved out to an encampment, making it possible once again for the public to have access to the models and records of the Patent Office. During their short stay, these troops managed to break 400 panes of glass; thus given ready access to the models which had formerly been shown behind glass, some of these soldiers stole patent models which caught their eye.[4] A different set of troops quartered in the Capitol Building managed to deface murals there.

Theodore Ruggles Timby (1822–1909) was an inventive genius from an early age. At the age of 16, he invented a floating dry dock and made a working model, but as he was too young to pursue it, he did not patent it, and it did

[3]See *Vindication of the Agricultural Division of the Patent Office* (1858), copy of pamphlet in National Archives Record Group 48 entry 266, box 1 of 4, [Bet63], and [Ver92].

[4]See [Edi61b] and [Edi61a].

not come into common use until later reinvented by others. When he was 20, he invented a revolving gun turret, made a working model, and filed a caveat in the Patent Office in 1843 to perfect his rights to file against subsequent inventors. In June of that year, he exhibited his model to President Tyler and his cabinet and in 1848 to a committee appointed to examine his plans, including Jefferson Davis. The committee then reported favorably, but nothing further was done at the time.[5]

Theodore Ruggles Timby

Early in 1861, the Confederate Navy began to raise the sunken Merrimac from Norfolk Navy Yard, intending to convert the ship to an ironclad. Reports of this reached Union authorities, causing some concern to Lincoln and his cabinet. About this time John Ericsson (1803–1889), a Swedish-born naval architect, engineer and inventor who had resided in America for 22 years, sent a proposal to President Lincoln offering to construct an ironclad vessel for the Union Navy. A naval board appointed to examine the situation recommended construction of Ericsson's vessel, and the keel of the vessel, named the Monitor by Ericsson, was laid on October 25, 1861. With almost unbelievable speed and with most of the plans existing only in Ericsson's head, the Monitor was launched 100 days later on January 30, 1862, and it was turned over to the government on February 19. The famous battle between the Monitor and the Merrimac took place on March 9, 1862.[6] A distinctive feature of the Monitor was its revolving gun turret. Theodore R. Timby filed patent applications to perfect his caveats, and he was granted two patents on the revolving turrets on July 8, 1862. See page 291 for the more important one. Although John Ericsson claimed that revolving gun turrets had existed before the nineteenth century, his associates bought Timby's patents almost immediately for an amount which Mr. Timby considered reasonable.

By June 1861, Judge Mason's business was slow, and he was feeling the depression and loneliness which beset him when he could not fight it with constant work.

[5]See Dictionary of American Biography 9:554–55 and [Par68].

[6]See Dictionary of American Biography 3:171–76.

On November 28, 1861, Judge Mason noted that he had visited the hospital in the Patent Office. These might have been casualties from the First Battle of Bull Run, fought over four months earlier, or possibly from the Battle of Ball's Bluff, a smaller battle fought only a month earlier. The north and west wings of the Patent Office were used as a hospital from October 1861 to January 15, 1863. About a year after Mason's visit, Munn & Co. reported that in two of the large model halls, cots had been placed in the passageways between the glass cases containing the models to provide accommodation for 800 patients. Yet it was said that the cots were so arranged that access might be easily had to the glass cases containing the models, so that the examination of inventions was not interrupted.[7] The picture of examiners, attorneys, inventors and the public wandering among the cots of the wounded and dying, searching for models relating to an invention, seems a shadow from another era.

John Ericsson

Clara Barton and Almira L. Fales, the wife of examiner Joseph T. Fales of Iowa, became nurses for wounded soldiers. Clara Barton had given her attention to the Massachusetts soldiers injured in the Baltimore riots on April 19, 1861, and she went immediately to the hospitals in Washington to care for wounded soldiers being brought in from First Bull Run. However, there were plenty of women in Washington willing to care for the wounded when they made it that far. She was appalled by stories of how poorly the wounded were treated in the field before they could be removed to the hospitals and of the lack of supplies for the treatment of wounded. She began advertising in New England newspapers for provisions for treating the wounded, and soon she had such abundant supplies that she established a distributing agency for the supplies. Later, she and Mrs. Fales followed the troops to the field and did nursing there. She was present on 16 battlefields and aided equally the wounded of both sides. Having discovered her life's work, she founded the American Red Cross after the war was over.[8]

In 1861, the first year of the Civil War, receipts of the Patent Office fell to $137,000 from $256,000 for 1860, while expenditures fell by only $31,000.

[7] See [Edi62].
[8] See [Bar22].

The inventors from 11 states were no longer filing applications, and, of those in the remaining states, many were off at war. The Patent Office was operating at a deficit. Commissioner Holloway dismissed five principal examiners, five assistant examiners, and five second assistant examiners. He also lowered the pay status of those examiners who remained. The cost of living in Washington increased because food could no longer be purchased from farmers in nearby Virginia. Congress had required the Patent Office to print specifications of issuing patents, but Holloway pled poverty of the Patent Office as his excuse for not doing so. Later in the war years, in 1864, Titian Peale wrote to his nephew that he could no longer afford his hobby of photography because, living on a fixed salary in a depreciated currency, with a greatly increased price of beef and bread and a doubled rent for his home, he no longer had sufficient money to buy photographic materials.[9]

On January 2, 1862, Charles Mason, Robert Washington Fenwick, and DeWitt Clinton Lawrence began business as the patent law firm of Mason, Fenwick and Lawrence of Washington, D.C. (This firm is, as of this writing in mid-1994, in the fourteenth decade of its existence, having recently merged with a larger firm.) Judge Mason soon noted that patent business was coming in more freely than it had been a few months earlier, although almost all inventions related to machines for war. He soon went to Matthew Brady's shop to have his photograph made. On March 29, 1862, the provost marshall took possession of the Church of the Ascension because the minister had not read a prayer the previous Sunday which the government had decreed was to be read in all churches. In March and April 1862, Judge Mason argued a reissue case for Silverthorn's patent against Charles M. Keller. In June 1862, Judge Mason was asked to represent Gail Borden (1801–1874) in his patent for preserving cider. Mr. Borden is better known for his evaporated milk.

Charges were made that Commissioner Holloway retained Southern sympathizers in the Patent Office and refused to investigate charges against them made by ardent Unionists. William C. Dodge was a patent examiner who went to see Mr. Holloway to demand that he dismiss draftsman Lewis Bosworth for disloyalty. Mr. Holloway thought the charges were unfounded and refused to dismiss Bosworth. Dodge then publicly and loudly denounced Bosworth as a traitor in the lobby of the Patent Office. By nightfall, the news of this charge was all over town, and acquaintances of Mr. Holloway were coming to him to ask if he were refusing to fire a traitor. Holloway called Dodge in to his office and asked if Dodge had personally seen Bosworth do anything that was

[9]See Titian R. Peale to Coleman Sellers 5/28/1864 Series VIII-A fiche 14 of 17, document 002171 Peale Family Papers [Kraus Microform], compiled by Smithsonian Institution, Washington, D.C.

disloyal. Dodge said that he only knew what people had told him. Holloway replied that he was not interested in hearsay. Dodge replied that by those standards neither of them knew that Jefferson Davis was disloyal, because neither had observed Jefferson Davis do anything disloyal. Holloway answered that unless better evidence could be provided than either of them could provide, Jefferson Davis could not be called upon to answer for anything. He then told Dodge that by making unsupported charges, he was making Holloway look bad, and that if Dodge continued Holloway would fire Dodge. Apparently, Dodge resigned just short of being fired, or perhaps he was actually fired. It was for reasons such as these that Mr. Holloway was investigated by a committee of the House of Representatives in February 1865 for refusing to fire known traitors.[10]

On July 1, 1862, the agricultural section of the Patent Office was separated from the Patent Office and formed into the Department (or Bureau) of Agriculture, with Isaac Newton appointed its first Commissioner.

Christopher M. Spencer

Christopher Miner Spencer (1833–1922) had conceived a design for a repeating rifle by 1857, and he built a wooden model by 1858. He perfected his invention by 1859 and filed a patent application, for which he was granted U.S. Patent No. 27,393 in March 1860. Some of these rifles were sold, apparently for use by the Navy, in the early stages of the Civil War. However, when General Stephen A. Hurlbut requested permission to arm his troops with the "Spencer Navy Rifle," Abraham Lincoln replied that he had personally inspected and fired one of these rifles and found it unsatisfactory. When news of Lincoln's decision reached the Spencer Repeating Rifle Company, the inventor went to Washington to demonstrate the rifle to the President. On August 18, 1862, Spencer went to the White House and explained his rifle to the President. He was invited back the next day for a test firing. When he returned the next day, Spencer and Lincoln, together with Lincoln's son Robert and a Navy officer, went over to the area around the Washington Monument to test the gun. Lincoln sent Robert to request Secretary of War Stanton to join them, but Stanton sent back a message that he was too busy. They set up as a

[10]See Congressional Information Services (38)HS-T.1 - T.6 microfiche.

target a board with a black spot painted near the end, and the President fired the seven cartridges in the rifle at the target. The first shot was low, but the last six were very close to the black spot. The Navy officer gave the board to Spencer as a souvenir.

The story that Lincoln then ordered 100,000 rifles as a result of the demonstration seems to be exaggerated, but it was apparently sufficient that the demonstration stopped Lincoln's objections to the desires of others to order the rifle.[11]

Walt Whitman

After the Seven Days Battle around Richmond, casualties in overwhelming numbers were brought into Washington. Five of the Episcopal churches in Washington were taken over for hospitals. Church services were held in Corcoran's picture gallery.

In August 1862, news came that there had been a second great battle at Bull Run and that it had been a victory but bloody. Aid for the wounded was called for, and Judge Mason went to Centerville in an all-night ambulance ride in the company of several physicians to do what could be done. But when they got there, they found that the Union had lost the battle, that the Confederates held the battlefield, and that there was nothing they could do. So they rode back to Washington in another all-night ambulance trip.

In January 1863, the poet Walt Whitman (1819–1892) was a volunteer in the hospitals around Washington, including the Patent Office Hospital. He wrote:

> The vast area of the second story of that noblest of Washington buildings, the Patent Office, is crowded close with rows of sick, badly wounded and dying soldiers. They are placed in three very large apartments. I have gone there many times. It is strange, solemn, and with all its features of suffering and death, a sort of fascinating sight. I go sometimes at night to soothe and relieve particular cases; some, I find, need a little cheering up and friendly consolation at that time, for they go to sleep better afterward. Two of the immense apartments are filled with high and ponderous

[11]See [MSD83] and Dictionary of American Biography 9:446–47.

glass cabinets, crowded with models in miniature of every kind of utensil, machine or invention it ever entered into the mind of man to conceive and with curiosities and foreign gifts.

Between these cabinets are lateral openings, perhaps eight feet wide and quite deep, and in these openings are placed many of the sick. Many of them are very bad cases, wounds and amputations. There is also a great long double row of them up and down through the middle of the hall. Then there is a gallery running above the hall in which there are beds also. It is, indeed, a curious scene, especially at night when lit up. The glass cabinets, the beds, the forms lying there, the gallery above, and the marble pavement under foot—the suffering, and the fortitude to bear it in various degrees—occasionally, from some, the groan that cannot be repressed—sometimes a poor fellow dying, with emaciated face and glassy eye, the nurse by his side, the doctor also there, but no friend, no relative—such are the sights in the Patent Office.[12]

Here is a case of a soldier I found among the crowded cots in the Patent Office. He likes to have some one to talk to, and I listen to him. He got badly hit in his leg and side at Fredericksburg that eventful Saturday, 13th of December. He lay the succeeding two days and nights helpless on the field, between the city and those grim terraces of batteries. His company and regiment had been compelled to leave him to his fate. To make matters worse, it happened he lay with his head slightly down hill, and could not help himself. At the end of some fifty hours he was brought off the field, with other wounded, under a flag of truce. I ask him how the rebels treated him as he lay during those two days and nights within reach of them—whether they came to him—whether they abused him? He answered that several of the rebels, soldiers and others, came to him at one time and another. A couple of them, who were together, spoke roughly and sarcastically, but nothing worse. One middle-aged man, however, who seemed to be moving around the field, among the dead and wounded, for benevolent purposes, came to him in a way he will never forget; he treated our soldier kindly, bound up his wounds, cheered him, gave him a couple of biscuits and a drink of whiskey and water, and asked him if he could eat some beef. This good secesh, however, did not change our soldier's position, for it might have caused blood to burst from the wounds, clotted and stagnated. Our soldier is from Pennsylvania. He has

[12]See Walt Whitman's newspaper dispatch for 2/23/1863, [SW63, pp. 39–40].

> had a pretty severe time; the wounds proved to be bad ones. But he retains a good heart, and is at present on the gain. (It is not uncommon for the men to remain on the field this way, one, two, or even four or five days.)[13]

There is also an account of Abraham Lincoln's visit to the Patent Office Hospital.

> One of the most unique hospitals in Washington was that organized in the museum of the Patent Office. Each alcove was a well-ventilated and lighted ward. The tessellated marble floors were covered here and there with clean matting, and the general aspect of the place was pure and neat. The President and Mrs. Lincoln, accompanied by Mrs. Doubleday (wife of General Abner Doubleday) and myself, were once visiting the Patent Office hospital, and the two ladies, being a little in advance, left us lingering by the cot of a wounded soldier. Just beyond us passed a well-dressed lady, evidently a stranger, who was distributing tracts. After she had gone, a patient picked up with languid hand the leaflet dropped upon his cot, and glancing at the title, began to laugh. When we reached him, the President said: 'My good fellow, that lady doubtless means you well, and it is hardly fair for you to laugh at her gift.' 'Well, Mr. President,' said the soldier, who recognized Mr. Lincoln, 'how can I help laughing a little. She has given me a tract on The Sin of Dancing, and both my legs are shot off.'[14]

In January 1864, Judge Mason began representing Clarissa Britain, a prolific female inventor, most of whose inventions related to domestic devices. On February 5, 1864, Mr. Mason, his wife, and a Mr. Lawrence of Virginia went to the theater to see Laura Keene in Our American Cousin, later famous as the play at which Lincoln was assassinated in 1865.

[13]See Walt Whitman's newspaper dispatch for 1/21/1863, [SW63, pp. 36–37].

[14]See [Bro95].

Patent Office Ball for Lincoln's Second Inauguration

Almost two years after the wounded soldiers had been removed from the Patent Office Hospital, the great hall of the Patent Office was used again, on March 6, 1865, this time for the second inaugural ball of Abraham Lincoln. Walt Whitman reported:

> I have been up to look at the dance and supper rooms for the inauguration ball at the Patent Office, and I could not help thinking what a different scene they presented to my view a while since, filled with a crowded mass of the worst wounded from the war, brought in from second Bull Run, from Antietam, and from Fredericksburg. Tonight, beautiful women, perfumes, the violins' sweetness, the polka and the waltz—then the amputation, the blue face, the groan, the glassy eye of the dying, the clotted rag, the odor of wounds and blood, and many a mother's son amid strangers, passing away untended there, (for the crowd of the badly hurt was great, and much for the nurse to do, and much for the surgeon.)[15]

[15]See Walt Whitman's newspaper dispatch for 3/6/1865, [SW63, p. 95].

That Monday evening, 4,000 guests came to the inaugural ball and banquet, climbing the curved double granite stairways from the south portico entrance, past the gas lamps that lit their way, and into the recently completed grand exhibition hall, then to the classic eastern hall (now the Lincoln Gallery) and from there into the elegant north saloon of the Patent Office. Gentlemen paid $10 each for themselves and the ladies they escorted, the proceeds going to the families of soldiers in the field. The ladies were magnificently dressed in silk and the gentlemen in formal evening attire, the court dress of foreign countries and dazzling uniform. Two bands played for dancing. At 10 in the evening, Lincoln entered the hall, escorted by Speaker Schuyler Colfax, followed by Mrs. Lincoln, escorted by Senator Charles Sumner. Shortly after midnight the guests went to the supper room where they had an extensive buffet dinner. It was a magnificent occasion to honor the President on his re-election and success in the nearly completed war. The President and First Lady left at 1:30 in the morning, but the dancing and eating lasted until about 4. Slightly more than a month later, the war was over and the President was dead.

Oddly enough, at a time so close to the end of the war, men were still paying large prices for substitutes to avoid being drafted. Judge Mason's office draftsman, Mr. Campbell, had just paid $900 for a substitute for one year when Mason noted it in his diary on March 12, 1865.

Chapter 27

The Confederate Patent Office

There is something about the American people that demands a patent system. The United States had a patent system before it had a thirteenth state. The Americans who founded the Republic of Texas demanded and got a patent system. It is not out of character that the Americans who seceded from the United States and formed the Confederate States of America demanded and got a patent system.

On April 29, 1861, Jefferson Davis sent a message to the Confederate Congress that the government was already receiving 70 applications a month, and he recommended that a Patent Office be set up promptly.[1] On May 21, 1861, even before the secession of Virginia and Tennessee from the United States, before the removal of the Confederate government from Montgomery, Alabama, to Richmond, Virginia, the Confederate government passed an act to establish a Patent Office. The principal difference from the U.S. patent law was that under the Confederate patent law, slaves could receive patents for their inventions.

Rufus R. Rhodes

As its Commissioner of Patents, the Confederate government chose Rufus Randolph Rhodes,

[1] See [Edi61e].

former United States patent examiner and member of the Board of Appeals, who had headed South when secession and war seemed inevitable. It is unclear precisely when the Confederate Patent Office opened for business. The appointment of Rufus Rhodes as Commissioner was announced in early July 1861, and Rhodes was said to have opened his office in Richmond by then and to soon be ready to proceed to business.[2] Confederate Patent No. 1 was issued on August 1, 1861. Since patents issued with some regularity immediately after that, it seems that the office opened for business about August 1.

C.S.A. Patent Office

The office occupied the whole third story of the Mechanic's Institute Building at the corner of Ninth and Franklin streets in central Richmond. A large hall occupying four-fifths of the space available on the floor was fitted out with cabinets with glass doors to allow exhibition and inspection of patented models.[3] The main hall of this building was the largest and best public hall in the city. For a time, the Secession Convention met in the hall. When the seat of the Confederate government was removed in July of 1861 from Montgomery to Richmond, those floors of the building which were not occupied by the Patent Office were occupied by the headquarters of the Confederate War Department.[4]

Not surprisingly, most of the patent applications received by the Confederate Patent Office related to devices to aid in the winning of the war. An example which was given in the first Annual Report of the Confederate Commissioner of Patents was that of Alfred G. Hearn, a village schoolmaster in Arkadelphia, Arkansas, who invented an instrument for measuring distances without the use of logarithms or other difficult process of calculation. For this he was granted Confederate patent No. 36, issued September 1, 1861. The inventor expected this to be of practical value in the adjustment of artillery to different ranges. At the time of writing, the invention was about to be tested with

[2]See [Edi61g].
[3]See [Rho63].
[4]See *Richmond Dispatch* 6/30/1896 p. 16.

guns at Nashville. In the Annual Report for 1862, Mr. Rhodes noted that the receipts of the Patent Office dropped dramatically in the second quarter, due to occupation of considerable portions of the Confederate States by the Union Army. Thus, the clerical force was reduced. Mr. Rhodes was proud that in each instance when a gentleman found that he had insufficient work at his desk to keep him occupied he resigned, and no dismissals were ever necessary. To the end of the existence of the Confederate Patent Office, in spite of heavy depreciation of the currency, the Patent Office paid its own way without necessity of any appropriation from the Confederate government. Through the end of 1864, as noted in the Annual Report for 1864 made in early 1865, the Confederate Patent Office issued 266 patents. It is believed that a few more patents were issued in the first few months of 1865. Toward the end, the only examiner in the Confederate Patent Office was Americus Featherman.[5]

Perhaps the most famous invention patented in the Confederate Patent Office was the subject of patent No. 100, granted to John Mercer Brooke for the design of the Confederate ironclad ship Merrimac, or, using its more proper but seldom used name, the C.S.S. Virginia. The claimed invention related particularly to the feature of that ship wherein the hull of the vessel extended underwater both fore and aft beyond the ends of an iron shield covering the above-water portion of the ship. See page 288 for a copy of this patent.

There is a story behind this patent. It seems that John Mercer Brooke, of the U.S. Navy—the same officer who escorted the Japanese diplomats to Washington in 1860—had submitted a plan for raising and reconstructing the sunken *U.S.S. Merrimac* as an ironclad warship at the request of Confederate Secretary of the Navy Stephen Russell Mallory. Naval constructor John L. Porter also worked on the project and made the first formal drawings for the project, as well as supervising the construction, though with frequent input from Lieutenant Brooke. Soon after the battle with the Monitor, the parties and their friends began arguing in the newspapers about who was the true inventor of the ironclad.

Brooke decided that the place to have the controversy decided was in the Confederate Patent Office, and thus he filed an application for a patent on his invention on May 2, 1862. His drawings were tracings of the drawings which Porter had made of the original plans. However, constructor Porter did not contest the patent application, and a patent on the invention was granted to John M. Brooke on July 29, 1862, as Confederate patent No. 100.[6]

[5]See Letter book B, Confederate Patent Office, Museum of the Confederacy, Richmond.
[6]See [Bro91].

John M. Brooke

The patent did not settle the matter, however, and the friends of Porter continued to maintain that he was the inventor up into the late 1880s. In 1891, Brooke, who had been a professor at Virginia Military Institute since immediately after the war, published a statement in a pamphlet, defending his claim. Because of this pamphlet, we have access to something usually otherwise unavailable, the specification of his Confederate patent

Near the end of the war, during the general evacuation of Richmond by Confederate forces, Confederate troops set fire to a warehouse to prevent the contents from being taken by approaching Union troops. The fire spread and on April 3, 1865, burned a large part of central Richmond, including the building in which the Patent Office was located. Apparently, all of the models were burned, although there is a suspicion that some of the written records may have been evacuated from the city. In June 1865, after the war was over, the United States Commissioner of Patents sent former Governor Leonard J. Farwell of Wisconsin, who was by then one of the principal examiners in the U.S. Patent Office, to Richmond with orders from high-ranking Army officers to support him, to recover any records of the Confederate Patent Office. He found nothing.[7]

Many years afterwards, two partially completed ledger books, which must have been only a small part of the total records, were delivered by private parties to the Museum of the Confederacy, located in Richmond. These, and a few original patents as delivered to patentees, plus the four Annual Reports of the Confederate Commissioner of Patents and a scattering of other printed reports, statutes, etc., constitute the entire known surviving records of the Confederate Patent Office.[8]

Confederate Patent Office Seal

On July 24, 1865, Rufus Rhodes passed through Washington and stopped to see

[7]See [Edi66a].

[8]See [New30] and [Tuc21].

Charles Mason. They had a long discussion
on conditions in the South. He then moved to New Orleans and opened a practice as an attorney at law and solicitor of patents. He remained in practice there until his death by apoplexy (stroke) in 1870 at the early age of 52.

During the years from 1861 through 1864, the Confederate Patent Office had issued 266 patents. A list of these patents is given in the appendix, beginning on page 273. During the same four years, the United States Patent Office issued 16,051 patents.

Chapter 28

Last of the Little Patent Office

David P. Holloway left office in mid-August 1865, and Thomas Clarke Theaker (1812–1883) was appointed Commissioner of Patents on August 15, 1865. Mr. Theaker had been a machinist and wheelwright, and was elected to Congress from Ohio, serving from 1859 to 1861. When he was not re-elected, he was appointed a member of the Board of Examiners-in-Chief (the Board of Appeals), and was in that office when appointed Commissioner.

The end of the war brought a surge of activity in the Patent Office. The number of applications filed in 1865 was 40 percent larger than in any preceding year. The number of patents granted was almost double that of 1861. The surplus money in the Patent Fund was increased during the year by more than 125 percent. It was a banner year for the Patent Office.

Thomas C. Theaker

Commissioner Theaker wrote to the Secretary of the Interior asking if citizens of the states recently in rebellion were entitled to take out patents as formerly. Secretary James Harlan consulted President Johnson, who instructed him to reply that each such resident must furnish an original or authenticated copy of the prescribed amnesty oath to the Patent Office before a patent could be issued, and if the applicant were in the "excluded classes," such as a former officer of the Confederate government, he also must provide evidence of a special pardon by the President.[1]

By 1866, the Commissioner was complaining in his Annual Report of a great shortage of room in the Patent Office Building for the Patent Office itself. More examiners were required to carry out the work assigned to the Patent Office, but the Commissioner thought there was no room for any more examiners in the available space. Every year from 1863 through 1867, the number of patents issued that year increased by between twenty and forty percent from the previous year.

Beginning in November 1866, the Patent Office began printing the specification of each patent as it issued. Previously, the Patent Office had to make two manuscript copies of each specification, one to affix to the issued patent and one for the record copy in the Patent Office. In 1861, when printing specifications was first tried and found to be too expensive, the printed copies were made in addition to the two manuscript copies. In 1866, Commissioner Theaker calculated that if printed copies of the specification were used for the issued patent and for the record copy, the expense of making the manuscript copies could be eliminated. The Patent Office contracted with Philp and Solomons of Washington for 10 printed copies of each specification at a cost said to be slightly less than the cost of two manuscript copies. The actual cost was 27 cents per 100 words for composing and printing 10 copies of the specification, compared to the usual cost of 10 cents per 100 words for manuscript copies, which does not seem to equal the numbers provided by Mr. Theaker. The remaining copies could be sold to the public at a lower price than formerly charged for manuscript copies, while still adding to the net income of the Patent Office. There seems to have been some opinion in Congress that the Public Printing Office should have done the work instead of a private contractor.[2]

[1]See James Harlan to Thomas C. Theaker 11/22/1865 National Archives Record Group 48 entry 266 box 1.

[2]See Thomas C. Theaker to O.H. Browning 2/16/1867 National Archives Record Group 48 entry 266 box 1.

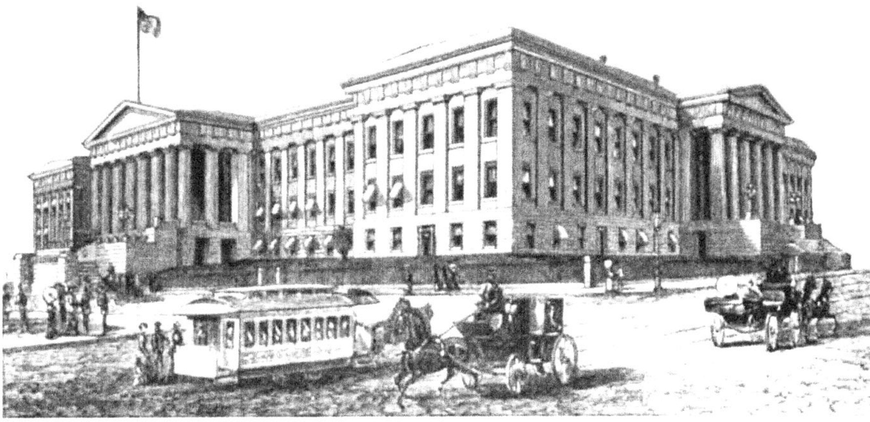

New Patent Office After Completion

In 1867, the fourth and final wing of the then New Patent Office Building was completed. However, the Interior Department preferred to think of the building as the Interior Department Building, and the Patent Office was not allowed enough space to accomplish its mission.

In 1867, the Patent Office issued 13,015 patents. For the first time, the Patent Office issued more patents in one year than the 10,000 it issued in the first 46 years and lost in the fire of 1836. In 1867, the Patent Office issued 10,000 patents in less than 46 weeks.

In late December 1867, Commissioner Theaker resigned.[3] According to Charles Mason's December 10, 1867, diary entry, Commissioner Theaker had just been given notice to quit. However, he remained in office until he submitted the Annual Report for 1867 on January 14, 1868, which was said to have been about his last official act.[4]

After Mr. Theaker's resignation, the Chief Clerk of the Patent Office, Gen. A.M. Stout, served as acting Commissioner from January 20 to July 24, 1868. Immediately upon his assumption of the office, Mr. Stout was asked why the three daughters of the late Chief Justice Roger Taney (1777–1864) were not receiving their fair share of the copying which the Patent Office had done. These copying jobs were much in demand by impoverished genteel ladies. Mr. Stout informed Secretary Browning that the name of one daughter had never been sent to the Patent Office to be given copying jobs, and thus she had never

[3] See [Edi67].

[4] See [Edi68b].

received any. The other two daughters each were being given about $50 per month in copying, which compared favorably with the other copyists, except those ladies having custody of assignment books.[5]

Examiners at Work

On April 23, 1868, Munn & Co. wrote to Secretary Browning to tell him that Scientific American had published all claims of all patents weekly for many years. In the past, the copies for Munn & Co. were made by Patent Office copyists at the rate of ten cents per word. Recently, the Patent Office had been issuing printed copies of the claims each week. Munn & Co. asked that the acting Commissioner be directed to give them copies free of charge, as they were furnished to foreign patent offices free. If they were to be charged, they wanted the charge to be no greater than the usual price for other printed matter which was sold by the Patent Office. Mr. Stout wrote to the Chief Clerk of the Interior Department, Gen. J. C. Cox, informing him that the only free copies were given to the examiners and to the British Patent Office. The British Patent Office, of course, gave the U.S. Patent Office copies of their entire issue of patents. The American Artisan paid the Patent Office $100 a week for a stereotype mat of the claims. No one else was given free copies. Mr. Stout wrote that the only possible question was whether printed copies of claims should be furnished at five cents per 100 words, as was done for printed patents, as opposed to 10 cents per 100 words, as was done for manuscript copies in general. The available record stops here.[6]

If it sounds like there were a lot of generals working for the Patent Office and throughout the Interior Department about this time, that is because there were. The war was over, and generals had to be recycled somehow. In late 1868, the examiners in charge of land conveyance were Gen. Albin F. Schoepf and Gen. William Henry Browne; Gen.— later Commissioner— Ellis Spear had been an examiner since 1865. For the next decade or two, Civil War generals occupied many of the high positions in the Patent Office.

[5]See A.M. Stout to Browning 1/20/1868 National Archives Record Group 48 entry 266 box 2.

[6]See Munn to Browning 4/23/1868; A.M. Stout to J.C.Cox 4/30/1868, both National Archives Record Group 48 entry 266 box 2.

On July 28, 1868, Elisha Foote (1809–1883) was appointed the eleventh Commissioner of Patents. He, like Theaker, had been a member of the Board of Examiners-in-Chief before his appointment as Commissioner. Mr. Foote had been a judge in western New York state and was later a patent attorney, inventor, and author of several books and papers on mathematics.

Elisha Foote

In his short term in office, Commissioner Foote accomplished several things. Until Foote changed the practice, it was common for application papers to be returned to the applicant for revision. The necessary changes would be written into the papers, or new papers would be provided, and there would be a new examination. Foote decided that papers once filed in the Patent Office must remain in the office files and could not be inspected by the applicant for any purpose whatever. For the first time, a patent would have a file history which could be examined after the patent issued. No changes could have been made in any papers filed in the Patent Office.[7] Commissioner Foote refused to pay the bills from Dempsy and O'Toole under a contract for furnishing stationery for the Patent Office as awarded by Commissioner Theaker, because he felt that the contract had been awarded because of fraud. Under this contract, bond paper was being purchased at eight cents per sheet. It seems that in evaluating competing bids, the Patent Office purchasing agent did not weight the various items according to how much was being purchased. The result was that a bid for \$8,000 was considered lower than a bid for \$3,000. Other defects were also found in the contract. When the Commissioner refused to pay the bills, Secretary of the Interior O. H. Browning appointed a committee to examine the matter, including examiners B. F. James, Norris Peters, and E. W. Griffin. When the committee was unable to discover anything deserving censure, a special committee of the House of Representatives was appointed and directed that all papers and correspondence relating to the matter be forwarded to them. Apparently, the Secretary attempted to make a farce of the investigation by limiting what the committee could inquire into, and by assur-

[7] See [Edi68a].

ing that they had no power to coerce witnesses to appear or to testify under oath.[8]

Mr. Foote appointed examiners W. B. Taylor, J. W. Jayne and examiner-in-chief B. F. James as a board to examine candidates for positions in the office. In addition all examiners currently in the Patent Office had to pass the same examination in order to keep their positions. Mr. Foote wrote that he intended to raise the standards of the principal examiners to that of the judges in the trial courts of the country, and to eliminate the practice of appointing persons as examiners just because they happened to be related to a congressman or to a senator.[9] There had been earlier employment exams in the Patent Office,[10] but this set of exams preceded the general Civil Service System examinations by about 15 years.[11]

Samuel Sparks Fisher

Mr. Foote left office about March 1869 and was replaced by Samuel Sparks Fisher (1832–1874) about the end of April 1869. Fisher had been a colonel in the Ohio National Guard during the Civil War, and was in practice as a patent attorney in Cincinnati before and after the war. He had been a precocious child, able to read a primer at age two. His father, a professor of chemistry, had encouraged his early learning.

Female copyists had been employed to work in their own homes for many years and were paid at 10 cents per 100 words copied. At the beginning of 1869, about 65 women were employed to work at home. An act was passed on March 3, 1869, authorizing 53 female copyists to be employed to work within the confines of the Patent Office Building, at a salary of $700 per year. Six rooms were set aside for their work when they began on July 1, and it was soon found that their work was fully equivalent in quality and quantity to that of the male clerks doing the same work at $900 per year. The number of male clerks was reduced, and Commissioner Foote recommended equal pay for equal work.[12]

[8]See [Edi69a] and [Edi69c].
[9]See [Edi68c].
[10]See [Edi61d].
[11]See [Edi84].
[12]See [otCoP69].

Congressman Thomas Allen Jenckes of Rhode Island (1818–1875) had sponsored two bills to establish of a department of civil service in 1868, with competitive examinations and initial probation, but he did not succeed in getting them passed.[13] Mr. Jenckes was a patent attorney who was the chairman of the House Committee on Patents about this time and managed to introduce several useful bills to improve the patent system and to get some of them passed. He introduced a bill in 1866 to pay examiners the money withheld from them when their salaries were reduced in the hard times at the beginning of the Civil War.[14] He introduced another bill in 1869 to appropriate money to enable the Patent Office to lithograph the drawings of issued patents. However, Congressman Henry L. Dawes managed to stop this bill, arguing that the Patent Office, although rapidly growing in importance, had been seriously crippled already by the thieves who were issuing its contracts on the one hand and by inconsiderate legislation on the other hand.[15]

Thomas Allen Jenckes

Mr. Jenckes made several proposals, in one or more bills, in the spring of 1870, to revise the patent system. He proposed to abandon publication of the Annual Report with its claims and drawings from all patents which had issued in the past year and to substitute the publication and weekly distribution of complete specifications and drawings from all patents issued that week. Photolithography had been developed at this time to the point where photographic copies of all patents could be printed at a very low cost. One estimate was that about 180 copies could be printed of 14,000 patents per year for about $100,000, or about four cents per copy. At the time, hand-drawn copies of the drawings from a patent, when required, cost from $10 to $100. A joint resolution supporting such publication and gratuitous distribution to each state was approved by Congress on January 11, 1871.[16] Mr. Jenckes also presided over the committee which pushed a bill through Congress in 1870 to thoroughly amend and codify the patent laws. When Mr. Ruggles put through his bill in 1836, a good law resulted, but it was a law for a country which had

[13]See National Cyclopedia of American Biography 8:436.

[14]See [Edi66b].

[15]See [Edi69b].

[16]See [Edi70c] and Samuel Duncan to TAJ 1/24/1871, manuscript papers of Thomas A. Jenckes, Library of Congress, Washington, D.C.

issued 10,000 patents in the past 46 years. Times had changed, the country was much more prosperous, and the Patent Office had issued about 14,000 patents during the most recent year. Mr. Ruggles revolutionized the patent system of his day; Jenckes refined that of his time. He improved appeals procedures, proposed but withdrew a requirement for maintenance fees, defined the time when rejected applications became abandoned and not subject to further prosecution, allowed a party sued by a patentee to win the suit by having the patent declared invalid, and even introduced laws to allow registration of trademarks in the Patent Office. He gave an impassioned speech in Congress on the necessity for patents which may have later secured the passage of his patent codification bill, which became law on July 8, 1870.[17]

Who was this Thomas Allen Jenckes? Remember Joseph Jenks Sr. (back in Chapter Two), who got the first patent for an invention in British colonial America. Thomas Allen Jenckes was his great-great-great-great-great grandson. His father changed the spelling of Jenks, but it was the same family.[18]

James A. Bland

Allen M. Bland (ca. 1833–poss. ca 1882) may have been the first black clerical employee of the Patent Office. Many conflicting stories are told about Bland. He has been said to have been the first black examiner, but your guide can find no record that he was ever an examiner. He was from Charleston, S.C., and attended a preparatory school at Oberlin College. It is also said that he graduated from Wilberforce University, but no record of his attendance at that school has been found. In 1854, he was living in Flushing, New York, and by 1860 he was a school teacher in Rensselaer County, New York. He came to Washington as a tailor in 1865, and was a clerk in the Patent Office by about 1870. He held this job until about 1874 while attending law school at Howard University. He was afterwards a tailor located near Howard University through 1882. He is known primarily as the father of James A. Bland (1854–1911), a musician, minstrel and author of Carry Me Back to Old Virginny, In the Evening by the Moonlight, and Oh, Dem Golden Slippers.[19]

[17]See [Edi70a] and [Edi70e].

[18]See National Cyclopedia of American Biography 8:436 and 27:221.

[19]See [Mil39], [LW82], [Spr80], and [Fle84].

Chapter 29

The Patent Office Begins to Look Modern

Samuel L. Clemens (Mark Twain) believed in the patent system. In his *A Connecticut Yankee in King Arthur's Court*, Hank Morgan, the Connecticut Yankee, said that "the very first official thing I did, in my administration—and it was on the very first day of it too—was to start a patent office; for I knew that a country without a patent office and good patent laws was just a crab and couldn't travel anyway but sideways and backwards."[1] Clemens had three patents on inventions of his own. His most famous patented invention was his self-pasting scrapbook, a book with adhesive already on the pages so that items could be pasted in by wetting the page to activate the adhesive. This scrapbook sold 25,000 copies, which he said "was well enough for a book that did not contain a single word that critics could praise or condemn."

But as an inventor and promoter of inventions, Twain could not equal Col. Beriah Sellers, a character in The Gilded Age, written in 1873 jointly by Mark Twain and Charles Dudley Warner. In Chapter 8, they wrote that the "Colonel's tongue was a magician's wand that turned dried apples into figs and water into wine as easily as it could turn a hovel into a palace and present poverty into imminent future riches." The Colonel was a confidence man except that he believed in what he was selling. At one point he says:

> I've been experimenting (to pass away the time) on a little preparation for curing sore eyes—a kind of decoction nine-tenths water and the other tenth drugs that don't cost more than a dollar a

[1] See [Twa89, Chapter 9].

barrel; I'm still experimenting; there's one ingredient wanted to perfect the thing, and somehow I can't just manage to hit upon the thing that's necessary, and I don't dare talk with a chemist, of course. But I'm progressing, and before many weeks I wager the country will ring with the fame of Beriah Sellers' Infallible Imperial Oriental Optic Liniment and Salvation for Sore Eyes—the Medical Wonder of the Age! Small bottles fifty cents, large ones a dollar. Average cost, five and seven cents for the two sizes. The first year sell, say, ten thousand bottles in Missouri, seven thousand in Iowa, three thousand in Arkansas, four thousand in Kentucky, six thousand in Illinois, and say twenty-five thousand in the rest of the country. Total, fifty-five thousand bottles; profit clear of all expenses, twenty thousand dollars at the very lowest calculation. All the capital needed to manufacture the first two thousand bottles—say a hundred and fifty dollars—then the money would begin to flow in. The second year, sales would reach 200,000 bottles—clear profit, say, $75,000—and in the meantime the great factory would be building in St. Louis, to cost, say, $100,000. The third year we could easily sell 1,000,000 bottles in the United States—profit at least $350,000—and then it would begin to be time to turn our attention toward the real idea of the business.

Mark Twain

In other words, Colonel Sellers was a typical optimistic inventor with an accounting system in place to account for his massive profits before he has finished the invention.

Strangely enough, the authors settled upon Beriah as the first name of Colonel Sellers only in the second edition. In the first edition, they called the Colonel by the name Eschol Sellers. Twain reported that "when the book had been out about a week, one of the stateliest and handsomest and most aristocratic white men that ever lived, called around, with the most formidable libel suit in his pocket that ever—well, in brief, we got his permission to suppress an edition of ten million copies of the book and change the name to 'Beriah Sellers' in future editions. The figure of ten million is taken from memory, and probably incorrect. Think it was more." And the gentleman who called? Of course. It was our friend George Escol Sellers.[2]

[2]See [TW73, Chapter 47].

Commissioner Fisher resigned from office as of November 10, 1870, and was replaced temporarily by Assistant Commissioner Samuel A. Duncan as acting Commissioner.[3] The first provision for an Assistant Commissioner was made by the consolidated Patent Act of 1870, and General Duncan was the first to occupy the position. When General Duncan left the Patent Office soon afterwards, he became the law partner of Colonel Fisher. Charles Mason and Scientific American both thought that the firm of Fisher and Duncan had undue influence with the Commissioner in 1874.[4]

In his only Annual Report, that for 1870, Mr. Duncan noted that the lithographing of drawings for patents being issued had begun. He also noted, as an item of expense, keeping a Patent Office horse. He recommended that the weekly listing of issued patents authorized by the consolidation Act of 1870 be expanded to be the only place of publication of all the advertisements that the Patent Office was required to make, and that it thus become the Official Gazette of the United States Patent Office. This was done on January 1, 1872. A person attempting to find such advertisements could thus find them all in one place. Mr. Duncan also noted with approval the recently begun publication of the Commissioner's Decisions.

On January 16, 1871, Mortimer D. Leggett (1821–1896) became the new Commissioner. Mr. Leggett was trained in both law and medicine and had been Superintendent of Schools in Zanesville, Ohio, before the Civil War. He had been a colonel heading the 78th Ohio volunteer regiment when General Grant had him promoted to brigadier general because of gallantry. He was later made a temporary (i.e. brevet) major general. When President Grant remembered Mr. Leggett and offered him a job in his administration, Leggett said that the only Government job that he wanted was Commissioner of Patents. Grant said that when the job became vacant, Leggett would have it, and he did.

Mortimer D. Leggett

In the Annual Report for 1871, Commissioner Leggett wrote that the Patent Office had upgraded the Patent Office pony by purchasing a horse, carriage

[3]See [Edi70d].

[4]See Mason diary 2/10/1870, manuscript papers of Charles Mason, Library of Congress, Washington, D.C., and [Edi75].

and livery. He noted the necessity for reproducing the drawings of previously issued patents to allow for proper management of the business of the office. By the time of the Annual Report of 1872, old patents were being printed rapidly. In some classes, all the old patents had been printed and could be consulted by examiners in their own rooms, while other examiners still had to go to the general portfolios and wait until examiners ahead of them were finished.

In January 1873, Titian Ramsay Peale, now age 72, was dismissed as an examiner because of alleged mental or bodily infirmity, and was appointed a class two clerk. Mrs. Peale maintained that he was as vigorous in mind and body as he ever was. There was no provision in the law at that time for a pension for long and faithful service to the government. Professor Henry of the Smithsonian protested that Mr. Peale's services were invaluable to the government. However, the Secretary of the Interior said that Commissioner Leggett did as he wished in the Patent Office. After 38 years of service to the U.S. government, Mr. Peale was to be offered a leave of absence as a class two clerk, in the absence of a pension. Later, Mr. Leggett extended his pay as examiner to the first of July 1873.[5]

E. T. Hall came in to see Commissioner Leggett with the original patent, issued July 31, 1790, to Samuel Hopkins. He offered to sell it to the government. General Leggett was anxious to secure it for the government but had no appropriate funds with which to purchase it. It belonged to Mr. Hall's elderly uncle, who was asking $500 for it.[6]

Chester Greenwood (1858–1937) was a 15-year-old lad living in Farmington, Maine, in 1873. Chester had large ears which stuck out from his head and formed excellent heat radiators. Unfortunately, in rural Maine in the winter, such ears were subject to frostbite and other painful treatment by the weather. In December 1873, Chester got a new pair of ice skates and went down to try them out on the recently frozen Sandy River. His enjoyment of his skates was cut short by the incipient frostbite of his ears. He went home and got some baling wire and bent it to form two loops at the ends of a foot-long piece of wire. He then asked his grandmother to sew a piece of fur over each of the loops. When she finished, he had the world's first earmuffs. Necessity is sometimes actually the mother of invention. Now Chester could skate. Other people saw Chester's ear protectors and wanted a pair of their own. Because the baling wire would not hold its shape well, Chester went on to design an adjustable steel band to hold the protectors over the ears. On March 13, 1877, 19-year-

[5]See Mrs. Titian Ramsay Peale to Anna Seward 6/4/1873 and Mortimer D. Leggett to Titian Ramsay Peale 6/17/1873, both Peale Family Papers [Kraus Microform], compiled by Smithsonian Institution, Washington, D.C., Series VIII-A fiche 15 of 17 .

[6]See E.T. Hall to W.A. Knapp 12/10/1879 National Archives Record Group 48 entry 266 box 4.

old Chester received U.S. Patent 188,292 for his ear mufflers. He went on to establish a factory in Farmington to manufacture earmuffs, and in its peak year this factory turned out 400,000 earmuffs a year. Nor was this the end of Chester's inventive career. He went on to obtain over 100 patents for his various inventions, including self-priming spark plugs, airplane shock absorbers, and the spring steel rake. He was one of the couple of dozen most prolific inventors in the nation, well behind Thomas Edison and *John Smith*, but a significant contributor to the progress and comfort of the nation nevertheless. In 1977, the State of Maine made December 21—approximately the winter solstice—an annual Chester Greenwood Day, to honor the man who made the inevitable cold weather of each upcoming winter just a little more bearable in Maine.[7]

Perpetual motion machines have a particular affinity for the Patent Office. The types of persons who propose such machines are sometimes astounding. The Reverend Doctor *John Smith*, or the same man by a different name, a local man well known at the Capital, and not suspected of studying any machinery but that of the moral law, appeared one day in the office of Commissioner Leggett.

Chester Greenwood

> "I know I've got it," he said.
>
> "What, sir?"
>
> "PERPETUAL MOTION, sir. Look!"

He set down a little machine. "If the floor were not in the way, if the earth were not in the way, that weight would never stop, and my machine would go on forever. I know this is original with me—that it has never occurred before to anybody else."

So enthusiastic was the doctor, it was with difficulty he could be restrained from depositing his money and leaving the experiment to be patented. So Commissioner Leggett quietly sent to the Patent Office Library for a book—a history of attempts to create perpetual motion. Opening the book, he pointed out to the astonished would-be inventor that his own machine had been tried and had failed more than a hundred years before. The reverend doctor took the

[7]See National Cyclopedia of American Biography 27:328–29; *N.Y. Times* 5/26/1977 C2:5; *Chicago Tribune* 2/16/1988 5:1; *Boston Globe* 12/22/1986 p 66; *Wall Street Journal* 2/10/1986 p 1+.

book out of the Commissioner's office, read, digested, and meditated thereon—then brought it back and placed it before the Commissioner in silence. No one ever heard him speak of perpetual motion again.

Another *John Smith*, this one an embalmer, sent a letter to Commissioner Leggett to describe a new process of embalming which he had originated. It was accompanied by an embalmed baby—submitted as the model which he requested should be placed in one of the glass cases of the Model Room. He considered himself deeply injured when his request was refused.[8]

John Marshall Thacher

John Marshall Thacher (1836–post 1895) of Vermont was appointed Commissioner of Patents on November 1874. He had been appointed to the Patent Office as a temporary clerk and soon as an assistant examiner in 1864. He had been an examiner-in-chief since 1870 and was Assistant Commissioner under Mr. Leggett. In his Annual Report for 1874, he noted that the Patent Office spent a large sum of money to lithograph the drawings from patents issued in previous years, but there was still a considerable sum of money received in excess of expenses. The specifications had been printed only since 1866, and he proposed to print all available specifications since 1836. Mr. Thacher resigned as Commissioner effective October 1, 1875.

Rodolphus (sometimes Robert) Holland Duell (1824–1891) of Cortland, New York, was appointed Commissioner on October 1, 1875. He had been a judge and had spent eight years in Congress. In his only Annual Report, that for 1875, he noted that there were only one-fourth of the old patents for which work had not at least begun toward reproducing their drawings. He pointed out that the office was so crowded that each of the 24 rooms assigned to principal examiners, each 20 by 20 feet, was occupied by between five to twelve persons.

On November 8, 1875, William Tell Steiger visited Commissioner Duell to return the volume that he borrowed from the Patent Office library just before fire in 1836, 39 years earlier. Commissioner Duell, recognizing an eyewitness when he saw one, asked Mr. Steiger to write out an account of his experiences in the early Patent Office. Steiger's explanation of the survival of the volume is that it had "escaped that conflagration by being accidentally in use at the

[8]See [Ame74].

writer's residence," but how a book can be accidentally in use is not clear to your guide.

In his letter, Steiger promoted an exaggerated version of the episode in 1814 when Dr. Thornton saved the Blodgett's Hotel from being burned by the British. He wrote:

R. Holland Duell

> ...during the war of 1812 with England, when the British captured the City of Washington and burnt the Capitol Building, a squad of soldiers with their officer, it is said, trained a loaded cannon, before the entrance of the Patent Office, with the intent of blowing it to atoms. When Doctor Thornton, the 'Superintendent' as he was then styled, threw himself before the gun, with the fires of seventy six flashing from his eyes, he demanded, 'Are you Englishmen, or only Goths and Vandals? This is the Patent Office, the depository of the ingenuity and invention of the American nation, in which the whole civilized world is interested. Would you destroy it? If so, fire away! and let the charge pass through my body!' The soldiers held down their heads, and the officer, much be it said to his credit ordered them and the gun away. And thus was the Patent Office saved on this occasion by the intervention of a very resolute man, the patriotic though eccentric Dr. Thornton. What if a similar stand had been taken at the burning of the Capitol?—that disgrace on the British name. Would it have stayed the match? Let us try to be charitable.

The volume returned by William Tell Steiger and the letter which he wrote were kept in the bookcase in the office of the Commissioner of Patents and Trademarks until sent out about 1990 for careful restoration as the only relic of the pre-1836 Patent Office.

Charles M. Keller died on October 15, 1874. Commissioner Leggett called a meeting of all of the members of the examining corps the next day, where he eulogized Keller as the Nestor of the Patent Office. Resolutions were passed expressing the examiners' high sense of the worth and talents of the man and of the loss of to the office and the profession by his death.[9]

[9]See *Official Gazette* 10/16/1874.

Chapter 30

The Second Patent Office Fire

By 1876, the Patent Office was so short of clerical help that it was unable to furnish certified copies of documents within several weeks of a request for copies, even when they were needed immediately in pending law suits. In the most urgent cases, attorneys were allowed to send their own clerks into the office to make the copies, while the Patent Office charged the same amount that it would have charged if its own clerks had done the work. By the end of 1876, the work of printing the drawings of prior and issuing patents was about 85 percent done. There were still 60,000 old patents whose specifications had not been printed at the end of 1876, and acting Commissioner William H. Doolittle earnestly recommended an appropriation to begin this work in the Annual Report for 1876.

The Centennial Exposition of 1876 in Philadelphia had attracted exhibits and visitors from all over the world. Alexander Graham Bell had made one of the first exhibits of his new telephone. This exposition marked a watershed point in the history of the nation, between a nation with an open frontier of unexploited land for settlers and a nation whose new frontiers were of unexploited technology for manufacturers and inventors. While the very exhibits of the Centennial Exposition were being set up in the exhibition halls, George Armstrong Custer's force was being killed by Indians on the Little Big Horn in Montana, and on the same day, 1,400 miles away in St. Louis, a double-header professional baseball game was being played. The Patent Office sent a carefully chosen 5,000 models of inventions for exhibit in Philadelphia. But, said Mr. Doolittle, the models were not needed to illustrate the value of the patent system. Every hall of the Exposition was filled with machinery and manufactured articles which illustrated the fruits of the patent clause of our Constitution far beyond the power of a few miniature models.

Ellis Spear

The next Commissioner was General Ellis Spear (1834–1917) of Maine, who was studying for the bar when the Civil War began. He raised a company of infantry which was assigned to the 20th Maine Infantry. At the end of the war he was a 31-year-old Brevet Brigadier General with an invalid wife and two little children. He cut the brass buttons from his military uniform, removed the shoulder straps and hat cord, and, thus attired in the best clothes he had, sought a civilian position as a patent examiner. He was hired as an examiner in 1865, was made Assistant Commissioner in 1874, and had resigned his office and practiced patent law for only two months when he was appointed Commissioner on January 29, 1877. After he left office, he practiced as a patent attorney in Washington until past the age of 80.

As 1877 dawned, the model rooms of the Patent Office had reached a zenith never before achieved and soon to be lost. The successors to the model rooms that Dr. Thornton had called the Museum of the Arts had become the National Gallery.

The Act of July 4, 1836, authorized the creation of a national gallery, and in the early days its future usefulness was recognized and every effort made to induce an exhibit of the manufacturing industries of the country therein. It is doubtful, however, whether its most enthusiastic early advocate ever anticipated the extent and diversity of its future contents.

South Hall of the Patent Office Model Room

The Model Room occupied the whole of the third floor of the Patent Office Building, immediately under the roof, and consisted of four grand halls, opening into each other, and affording a promenade of about one-fourth of a mile around the four sides of a quadrangle. These magnificent halls were fitted up with tiers of cases, the room being sufficiently high for two tiers, one above the other. Each case was eight feet in height by from sixteen to twenty feet in length. The cases were made of white pine, with glass sides and ends. They were so placed that there was sufficient room around each case to give easy access to them both for the casual visitor and for inventors and examiners. The cases could be opened and their contents inspected at any time in the immediate presence of an employee of the Patent Office. This great gallery was visited yearly by thousands of people, both for profit and pleasure. It contained about 200,000 models of American inventions, besides many curiosities and mementoes, specimens of home manufacture, and priceless treasures of deep historic interest.[1]

North Hall of Model Room (Rebuilt after Fire)

A guide book issued shortly before this time described the experience of visiting the museum.[2] The Patent Office Building occupied two undivided city blocks, fronting south on F Street, north on G Street, east on 7th Street West, and west on 9th Street West. The length of the building, from Seventh to Ninth Street, was 410 feet, and the width, from F Street to G Street, was 275 feet. It was built along all the four sides, with a large interior quadrangle about 265 by 135 feet in size. It was constructed in the plainest Doric style, of massive crystallized marble, and though devoid of exterior ornament it was one of the most magnificent buildings in the city. It was ornamented with massive

[1]See [Off77].
[2]See [Ell69].

porticoes, one on each front, which added much to its appearance. The eastern portico was much admired. That on the south front was an exact copy of the portico on the Pantheon at Rome.

The interior was divided into three stories. The ground and second floors were arranged as offices for the accommodation of the business of the Interior Department, including but not limited to the Patent Office. The third floor was occupied by an immense saloon or public exhibition hall, extending entirely around the quadrangle and used as the model room. Its south hall was a museum. The models and other articles were arranged in glass cases on each side of the room, ample space being left in the center for promenading. There were two rows of cases, one above the other—the upper row being placed in a handsome light gallery of iron, reached by iron stairways, and extending from both sides of the south hall entirely around the east, north, and west halls. The halls themselves were paved with tiles. The ceiling was supported by a double row of pillars, which also supported the galleries. The walls and ceiling were finished in marble panels and frescoes. The guide book said that a more beautiful saloon was not to be found in America.

The visitor entered the building from a beautiful south portico, passed through the marble hall, and up broad stairs to the door of the saloon. Entering it, the visitor found a large register, with pens and ink, at the right of the door, in which the visitor was expected to record name and date of visit. High up on the walls of this hall were four cameo-like bas-relief figures representing Benjamin Franklin, Thomas Jefferson, Eli Whitney and Robert Fulton. Keeping in mind the accusation made by Dr. Thornton back in Chapter Nine that Robert Fulton did not actually invent anything, one commentator said that the figures represented three inventors, two patentees, and one man who was both.

The first case to the right of the entrance contained Benjamin Franklin's printing press, at which he worked when a journeyman printer in London. It was old and worm-eaten, was only held together by means of bolts and iron plates, and bore but little resemblance to the steam presses of 1876.

Much of the remaining space in the south hall was occupied by exhibits on fire escapes and fire extinguishing, as well as other subjects. Several other cases contained original treaties of the United States with foreign governments. The treaties were written upon heavy vellum, in wretchedly bad hands, with a worn and faded appearance. All except the treaties with England and the Eastern nations were written in French, and all were furnished with a multiplicity of red and green seals which attracted the attention of the visitor. Among these was the first treaty of alliance with France—the famous Treaty of 1778—which gave the aid of the French king to the cause of the struggling states during the American Revolution.

Next to be seen were the relics of George Washington, considered to be among the greatest treasures of the nation. They included the camp equipment and other articles used by General Washington during the Revolution. Here were the tents which constituted his field headquarters, together with his blankets, the bed-curtain sewn for him by his wife, his window curtains, and chairs, his wash stand and mess chest.

Also to be seen was the coat worn by Andrew Jackson at the Battle of New Orleans, and the panels from the state coach of President Washington. Also on exhibit was the sword, formerly ornamented with diamonds, presented to Commodore Biddle. This sword was among the items stolen from the Patent Office in 1848 and afterwards recovered.[3]

Up until 1876, in an ordinary picture frame near the Washington relics was the original Declaration of Independence with the signatures of the members of the Continental Congress attached. Even in 1876, it was old and yellow, with the ink fading from the paper. In 1876, it was transferred from the National Gallery to the State Department.[4] Nearby hung Washington's commission as Commander-in-Chief of the American Army, bearing the bold, massive signature of John Hancock, President of the Continental Congress.

Close by was one of the most famous of all the patent models, that of Abraham Lincoln's invention. It was a plain model, roughly executed, representing the framework of the hull of a Western steamboat. Beneath the keel was a false bottom, provided with bellows and air-bags. The ticket upon it bore the memorandum: "Model for Sinking and raising boats by bellows below. A. LINCOLN May 30, 1849." Upon his inauguration, Mr. Lincoln got one of the employees of the Patent Office to find the model for him. After his death, it was placed in the Washington case. The opposite case contained the hat worn by Lincoln on the night of his assassination.

A few cases down the way, the visitor came upon a number of handsome silk robes and Japanese articles of various kinds, presented to President Buchanan by the team of Japanese ambassadors in 1860.

The remainder of the south hall and all of the three other halls were devoted to models of patented machinery and other inventions. The cases on both the upper and lower levels of the exhibit hall were well filled. There were models of bridges spanning the spaces between the upper cases, and models of large machines were laid on the floor of the hall. Here was everything the mind of the day could think of. Models of improved arms, clocks, telegraphs, burglar

[3]See [Kei74].

[4]See [Buc90].

and fire alarms, musical instruments, light-houses, street cars, lamps, stoves, ranges, furnaces, peat and fuel machines, brick and tile machines, sewing machines, power looms, paper-making machinery, knitting machines, machines for making cloth, hats, spool-cotton, for working up hemp, harbor cleaners, patent hooks-and-eyes, buttons, umbrella and cane handles, fluting machines, trusses, medical instruments of gutta percha, corsets, ambulances and other military equipment, arrangements for excluding the dust and smoke from railroad cars, railroad and steamboat machinery, agricultural and domestic machinery of all kinds, and hundreds of other inventions, lined both sides of the three immense halls. One could have spent a year examining them, and still have learned something new every day. For every article one could think of, there were at least half a dozen models, and there were many inventions to be seen of which nine people out of ten had never dreamed before. The number was increasing every year. As the country grew greater, new wants were felt. They were sure to be supplied, and the model room of the Patent Office kept a faithful record of the history of American civilization.

Supposedly, the Patent Office Building was nearly fireproof, but the events of September 24, 1877, demonstrated that this was not true. The original wings of the Patent Office had a roof with a framework of iron and brick, but by the time the west or Ninth Street wing was built, cheaper methods of construction had been allowed. The west wing roof was formed of trusses made partly of iron and partly of wood. These were covered with pine sheathing with thin copper sheeting over it. Over 100,000 board feet of white pine lumber was used in the construction of the roof of the west wing. Under the south end of the west wing was a conservatory or hot house, and just to the north of this some 12,000 rejected models were stored in the loft up among the trusses of the roof. A finer lot of kindling wood could not be assembled. The precise origin of the resulting fire was never determined with certainty, but a leading theory was that a defective flue passing through the section of the loft containing the rejected models set fire to the models or to the adjacent wooden roof.

About 11 o'clock on the morning of Monday, September 24, 1877, employees in the building discovered that dense clouds of smoke were issuing from the skylights of the building. Firemen were summoned immediately, but there was some delay in getting water to the fire because the fire was 80 feet above the street, and 20 feet above the water pressure in the hydrants. There was a tank of water outside the hot house, but that could not be reached because of the flames. By the time the pumping engines were on hand, the fire had spread seriously, sweeping over the entire west wing and bursting through the windows and portions of the roof. The first half hour of the fire was so serious that telegrams were sent to Alexandria and Baltimore, requesting that their engines also be sent. The one engine available in Alexandria was sent promptly, and four engines from Baltimore were sent down by rail and drawn by horses to the fire.

Clerks Passing out Models from the West Hall

Inside the Patent Office, employees were removing the books and papers from the most exposed offices, saving what they could. In the drafting division, live coals and molten metal were coming down the ventilator shafts. One employee stopped one shaft with a coal scuttle, and another was stopped with a water cooler, while the employees worked to remove 777 folios of drawings containing 211,243 original drawings. These were all removed without the loss of a single drawing from those stored in their folios. The original drawings from the issue of September 4, some 300 in number, were in the model room for use in identifying models and were entirely lost, but they had been reproduced by photolithography in the printed patents, and nothing of real value was lost in that lot. The drawings from 31 older patents in the field of woodworking were in the drafting division being traced for use in printing old patents and were lost to the fire.

Bringing Models down the Main Stairway

The Patent Office would attempt to do replacement drawings from the specifications, the only record surviving. Two of those 31 drawings were from patents which had issued before July 4, 1836, and had been restored once before. Most of the 114,000 models in the west and north halls of the model room were totally destroyed or at least seriously damaged. The 12,000 rejected models were all destroyed, but this was no loss—they had been kept only because, until recently, the Patent Office had been required to keep them. About 150 copies of the photolithographed drawings of 40,000 patents were destroyed by fire and water, but could easily be replaced by reproducing them from the few copies surviving in other locations.

During the fire, the corridors were crowded with people working desperately to save property. Books, papers, office furniture and models were carried out of the third floor to line the corridors of the lower floors. Just after the fire began, a brisk breeze from the south sprang up, so that the south hall, containing the valuable collection of historical objects, was not harmed. By mid-afternoon the fire was out, and the long clean-up effort began.

For months after the fire, intensive efforts were made to repair and restore as many of the 114,000 models as possible, and apparently about 27,000 were restored, leaving 87,000 models lost to the fire. Most of the work of restoring models was done in the north hall, the one that was formerly considered the finest of the four halls forming the model museum. It was the last wing of the Patent Office finished and was used as a hospital during the war. Just before it was fitted up to receive models, it was used for the Inaugural Ball at Lincoln's second inauguration, so that it had seen happy times as well as sad ones. By March 1878, it presented the appearance of a huge machine shop, except that comparatively little machinery was employed, which consisted mostly of small lathes run by foot power and two or three portable forges. The models were first picked out of what would appear to be heaps of scraps, and then were arranged as nearly as possible in the classes to which they belonged, the location in which they were found being in many cases the only clue to the class. The examiners in each class then compared the models with the drawings which accompanied them when originally filed, and they affixed a card to each giving the name of the inventor, date of the patent, and the name of the invention. The model was then entered in a book, with a description of its appearance and condition, and was then passed to the laborers for cleaning. The first operation was to pickle it in a solution of sulfuric acid to eat out the rust and dirt and then to wash it in a bath of lime water to counteract the acid left from the pickling tank, after which it was dried with sawdust. Then, if needed, the model was soaked in a bath of kerosene to loosen the screws and other adhering parts which could not otherwise be readily started. After draining, it was passed to a machinist, who then cleaned and repaired it as far as possible or allowable. In many cases the model had simply been bent out of shape by the heat, and it was separated into pieces and the bent parts straightened using portable forges. If any part was missing, a search was made for it among the miscellaneous mass of pieces, and when the missing part was found, it was replaced in proper position. In many cases, small parts were made and added to the model to make it complete, which parts, however, were always made to correspond exactly with the drawing. The model was then taken back to the bookkeeper, who entered it in his register a second time with a description of the part that had been added, and the model was then transferred to temporary cases in the West Hall, looking in many cases better than it did when originally deposited.[5]

The monetary loss from the fire of 1877 was many times that of the 1836 fire, but the information loss was only partial. Models were lost in great number, but the drawings corresponding to the models were unharmed. Some 600,000 photolithographed drawings were lost, but these were easily replaced merely by the expense of printing them again. Also lost were 300 original drawings which had already been printed. The major loss was of the drawings and the corresponding models for 31 older patents, leaving only the specifications from which to restore the patent. No patents were totally lost in the fire of 1877.

Sorting and Repairing Broken Patent Models

[5]See [Edi78].

Chapter 31

The End of the First Century of the Patent Office

Printing patents had an interesting and lasting effect on the furniture which was required by the Patent Office. Before patent drawings were printed, examiners, while searching patents, had to look at the original drawings in the draftsman's office in large portfolio cases, such as those shown in the illustration from 1869 on page 230. There were 777 folio cases of original drawings which were safely removed from the draftsman's office in the fire of 1877. But when the drawings, and especially the entire patents, were printed, they were available for search in the examiners' rooms and soon in the new public search room. The necessary new filing system was provided by the shoe cases in use in the 1994 Patent and Trademark Office. The origin of the term shoe is lost, although every patent examiner knows where his shoes are. Some have attributed the term to Thomas Jefferson, suggesting that he stored his patents in shoe boxes. But we know that the drawings kept in the Patent Office in the preprinting days were of varying sizes and were kept in portfolios from the earliest days. Shoes, as we know them, could only have arisen after all patents were available in small, uniform sizes and could be kept in small, uniform boxes. A complete inventory of the moveable property in the Patent Office was made in 1870,[1] including numerous portfolio cases, portfolio racks, portfolio drawers, and cases

[1]See Samuel S. Fisher to C. Delano 11/9/1870 National Archives Record Group 48 entry 266 box 2.

for models, 22,000 volumes of books and 300 spittoons, but no shoes. Augustus Burgdorf, livery stable operator, undertaker and cabinet-maker of Washington, sold portfolios and cases to the Patent Office in 1878.[2] The first known mention of shoes was on March 28, 1879, when he sold shoe drawers to the Patent Office for $115.[3] What were called shoe drawers would now be called shoes. Among the many possible origins of the term is that preferred by your guide. Perhaps file cabinets suitable for holding bundles of patents while allowing easy access to search through them were already available before they were needed. Perhaps shoe shops of the day kept their supply of ready-made shoes in wood cabinets containing numerous drawers of just the right size to hold patents, and when the Patent Office wanted to order its first drawers, it ordered shoe drawers from Augustus Burgdorf. Or maybe not.

Commissioner Spear requested permission in 1877 to reduce expenses by disposing of the horse and carriage which belonged to the Patent Office.[4] However, he pointed out, although the expense of $100 per month for the horse and carriage was paid by the Patent Office, it was not maintained exclusively for the use of the Patent Office, so he thought it proper to submit the matter to the Secretary of the Interior for decision. The reply is not available, but by 1878 the Patent Office spent less than $100 for the year in hiring horses and carriages from James Keleher as needed.

Halbert E. Paine

The next Patent Commissioner was Halbert Eleazer Paine (1826–1905) of Wisconsin. He graduated from Western Reserve at age 19 at the head of his class and practiced law in Cleveland, Ohio. He was brevet major general with the troops of Wisconsin in the Civil War and was later elected to Congress from Milwaukee. He was appointed Commissioner on November 1, 1878. He promoted the use of typewriters in the Patent Office, and he installed the first typewriter in his own office, with Virginia Middleton being his typist. He then issued an order for the general use of typewriters in the Patent Office, and had them installed throughout the Patent Office as fast as typists could be found. In the various records in the National Archives, the first

[2]See [otCoP78].

[3]See [otCoP79].

[4]See Ellis Spear to Carl Shurz 11/2/1877 National Archives Record Group 48 entry 266 box 2.

typewritten Patent Office document found by your guide was dated May 28, 1880, less than a month after Mr. Paine's departure from office.

Back in Chapter 29, E. T. Hall came to see Commissioner Leggett, offering to sell to the Patent Office the Hopkins patent, the first issued by the U. S. government. In 1879, Commissioner Paine wrote to the Interior Secretary,[5] recommending that Congress be asked to appropriate $500 to purchase the patent. The government did not purchase this document, and it eventually found its way into the collections of the Chicago Historical Society.

Edgar M. Marble

In 1879, the photolithographing of patent drawings was being done under appropriations which authorized the Commissioner to require the contractor to do the work in Washington. Earlier contracts had been let to firms in New York City, and excessive delay was found to result. The Patent Office was not happy having its only copy of various drawings out of the Patent Office, and it was exceedingly unhappy having it out of the city. Naturally, there were protests from distant lithographers that they were being improperly excluded from lucrative government contracts. The Patent Office maintained that it wanted the proof sheets within one and a half days from delivery of the drawings to the contractor, and that could not be done out of town. Delay in obtaining proofs would result in delay of issuance of the Official Gazette, which was not considered acceptable. Attempts were made twice to have the work done outside of Washington. The first time, there was a delay of two to three months in returning drawings to the Patent Office.

The second time, the work was considered generally inferior, and the proof sheets, even when inferior, were received in an average of five days. Both Mr. Paine and his Chief Clerk vigorously opposed attempts to change the terms of the appropriations. Congress decided the Commissioner should be allowed to have the work done anywhere he could have it done at reasonable rates.[6]

The next Commissioner was Edgar M. Marble (1838–1908) of Michigan, who was appointed May 7, 1880. In late 1881, he resigned to become land commissioner of the Northern Pacific Railroad in Duluth, Michigan. However,

[5]See Halbert E. Paine to Sec Int 12/19/1879 National Archives Record Group entry 266 box 4.

[6]See [Ano86].

Michigan in winter did not suit his health, and he returned to the Patent Office within weeks and remained until 1883.[7]

Martin Van Buren Montgomery

Next was Benjamin Butterworth (1837–1898) of Ohio, who was appointed Commissioner on October 26, 1883. He was a Republican member of Congress from Cincinnati from 1878 through 1882. When he was defeated for the 48th Congress, he was made Commissioner of Patents. He was of the impression, soon after assuming the duties of Commissioner, that the principal handicap under which the Patent Office operated was being caused for it by Congress. The Patent Office had insufficient examiners, insufficient room, and insufficient compensation to do the job as well as it should be done, and all of these were correctable by Congress without spending funds in excess of the fees already being paid by applicants. But Congress would not appropriate the money needed to do the job. While Commissioner, he compiled a book entitled The Growth of Industrial Art, an excellent collection of engraved illustrations of the various stages of development of many arts. Mr. Butterworth was reelected to Congress from Cincinnati for the 49th Congress and served in Congress from March 1885 through March 1891, serving eventually as the chairman of the House Committee on Patents. He refused further nomination to Congress. Then he had a unique accomplishment. He was appointed a second time to be Commissioner of Patents on April 12, 1897, and served until his death from pneumonia in early 1898.

Martin Van Buren Montgomery (1840–1898) of Michigan was appointed Commissioner on March 21, 1885. He was a lawyer practicing in Lansing, Michigan when appointed. He emphasized orderly and uniform practice in the Patent Office. Upon his resignation as Commissioner in 1887, he was appointed Associate Judge of the Supreme Court of the District of Columbia.

Benton J. Hall (1835–1894) of Iowa was appointed Commissioner on April 7, 1887. He was a Democratic member of Congress from Iowa from 1885 to 1887, and upon defeat for re-election he was appointed Commissioner. While he was Commissioner, he heard one of the most important interference cases argued by one of the greatest collections of legal talent ever to argue a case before the Patent Office. This was McDonough v. Gray v. Bell v. Edison,

[7]See [Edi81].

a continuation of the telephone interference cases. After leaving the Patent Office, he practiced patent law in Chicago.

Charles Elliott Mitchell (1837–1911) of Connecticut was appointed Commissioner on April 1, 1889. He was a prominent patent practitioner for years before his appointment. Mr. Mitchell was Commissioner during the Patent Office Centennial. When he left office, he practiced patent law in New York City for some years, then became president of the Stanley Rule and Level Company of New Britain, Connecticut.

Throughout the 1880s, successive Commissioners complained of the need for more room. When the Indian Bureau moved out of the Patent Office Building in 1884, nine additional rooms were acquired for use by the Patent Office. The walls of these rooms were covered with green mold. The rooms were damp, and the atmosphere in them was constantly foul. Several deaths were said to have occurred because of disease contracted in these foul rooms.

In 1878, the Patent Office had problems collecting money for some of the services it rendered. When a patent agent wrote to the Patent Office to ask for a handwritten copy of some patent specification, he was never certain what that copy should cost. He could ask for an estimate, then await the estimate before ordering the copy with the necessary funds, then await the copy, but this caused unnecessary delays. The Patent Office sometimes made copies and billed the patent agent for the amount due, but bankruptcies and deaths sometimes left the Patent Office unable to collect the small sums of money due it. Commissioner Spear then authorized the Patent Office financial clerk to receive money deposited in advance by persons who frequently ordered services, and then to spend this money on instructions from the depositors. The financial clerk kept this money in a collection of envelopes, which were marked with the name of the depositor and the sum on deposit. As each sum was collected from the depositor, the amount so collected was written across the front of the envelope. This was in effect the first deposit account. In November 1878, Levi Bacon (ca 1820–1887) of Michigan was appointed financial clerk of the Patent Office through the influence of Jay Hubbell, a Republican politician, also of Michigan. Bacon had been in the Patent Office in some capacity since September 30, 1875. He seemed to do his work so efficiently that he was held over by the subsequent Democratic administration. When Bacon died, still in office, on June 22, 1887, an audit of his accounts found shortages, and his deposit account envelopes were all found empty. The total shortage in his accounts was about $33,000, of which about $4,500 came from the attorneys' deposits. In those days it was common for the political party in power to make assessments of money against the politically appointed employees of the various departments, the money to be used to assure reelection of the party that had appointed them. A memorandum found in Bacon's safe indicated that the total amount of the shortfall which could not be accounted for by due-bills from employees and others, about $11,000, was very nearly the exact amount that Mr. Bacon had to contribute to the Republican coffers in 1880 as the assessment of the Interior Department.[8]

Benton J. Hall

There was much discussion about whether the Patent Office was to be responsible for the losses of the agents from their deposit accounts. Although it is not clear, it seems that the Commissioner got permission for the Patent Office to fill copying orders to the amount of the deposited money, though the money itself was missing. The long-term solution was to require that money deposited by agents be immediately deposited in the Treasury, with only deposit receipts kept in the clerk's envelopes for accounting purposes.

Although the consolidated Patent Act of 1870 had declared that it was only necessary to submit models with patent applications when required by the Commissioner, they were usually required for the next ten years. About 1880 the Commissioner issued a new regulation which required a model only when specifically required by the Commissioner or a principal examiner, no model being required if a working model were exhibited in operation in Washington to aid in the examination of the application.[9] In 1890, when 26,000 patents were issued, only 535 models were added to the collections in the Model Rooms. Commissioner Mitchell regarded it as a public calamity in 1890 that the Patent Office had a few years earlier been compelled to suspend the reception of models except in special instances, for want of room to store and exhibit them.[10] In the spring of 1883, Herman Hollerith (1860–1929), formerly a census clerk and an M.I.T. instructor, became an assistant patent examiner. He had been working on a method of tabulating census results with punched cards. Within a year he resigned from the Patent

[8]See *Washington Post* 6/23/1887 p. 3, 7/3/1887 p. 5; *National Republican* 7/2/1887 p. 1; *Washington Sentinel* 7/9/1887 p. 1, [otCoP87], and [otCoP88].

[9]See [Edi80a].

[10]See [otCoP90].

Office and opened an office as an "Expert and Solicitor of Patents," then applied for patents on his punched card machines. The patents were granted, and the cards were successful and were adopted by the Census Bureau for the 1890 census. Hollerith founded the Tabulating Machine Company, a predecessor to International Business Machines or IBM. He began the practice, later adopted by IBM, of renting, not selling, his tabulating machines.[11]

Korekiyo Takahashi

In early 1886, the Japanese government sent Korekiyo Takahashi (1854–1936), soon to be their first Commissioner of Patents, to Washington to study the U.S. Patent Office in an effort to learn how to set up the Japanese Patent Office. Mr. Takahashi had studied for a year in the United States when he was thirteen, and he was fluent in English. Mr. Takahashi was in Washington for several months, working diligently. The U.S. Commissioner had directed that every available facility was to be given him in collecting information, and that a copy of every publication, notice, circular, etc., ever issued by the Patent Office and then available was to be given him. Among the many patent examiners who had discussions with Mr. Takahashi was the design examiner, Dr. P.B. Pierce, who had several days of discussions with him. Near the end of these discussions, Dr. Pierce said: "Mr. Takahashi, I have answered many questions asked by you; would you object to answering a single question which I would like to put to you?" The Japanese Commissioner at once indicated his readiness to answer. Dr. Pierce said: "I would like to know why it is that the people of Japan desire to have a patent system." "I will tell you, then," said Mr. Takahashi. "You know that it is only since Commodore Perry, in 1854, opened the ports of Japan to foreign commerce that the Japanese have been trying to become a great nation, like other nations of the earth, and we have looked about us to see what nations are the greatest, so that we could be like them; and we said 'There is the United States, not much more than a hundred years old, and America was discovered by Columbus yet four hundred years ago;' and we said 'What is it that makes the United States such a great nation?' and we investigated and we found that it was patents, and we will

[11]See [Aus82].

have patents." Dr. Pierce, reporting the interview, added: "Not in all history is there an instance of such unbiased testimony to the value and worth of the patent system as practiced in the United States." As the present generation knows, Japan soon got its patent system.[12] In 1921, Korekiyo Takahashi was prime minister of Japan for seven months. He advocated economic instead of military competition with the Western powers, and for this was assassinated in 1936 by a group of young officers to get him out of the way of the military faction preparing for World War II.

W.H. Thorne

From at least 1887 through 1893, Mr. W. H. Thorne, formerly a second-rate actor from a family of first-rate actors from New York, was a copying clerk in the Patent Office. He claimed to have inherited the title Duke of Normandy from his grandfather, John Mastayer, and his friends all called him Duke. This peerage was first confiscated by the French court of peers in 1203 from John Lackland, later King John of England, for having assassinated his nephew Arthur, Duke of Brittany. It was re-established three times, each time being given to the eldest son of the King of France and each time being extinguished upon that son's assumption of the French throne, the last time in 1469. The title had been extinct for over 400 years when it was claimed by Mr. Thorne. But every second-rate actor needs something to make other people feel that he is important.[13]

The last entry in the Annual Reports up to 1892 for the expense of keeping a horse for the Patent Office was made in 1887. After that, nothing.

As early as 1887, F.M. Shields of Mississippi wrote a letter to *Scientific American*, suggesting that inventors meet in 1890 for a celebration of the centennial of the U.S. patent system.[14] On April 8–10, 1891, the patent system celebrated the beginning of its second century. The celebration was a year late, since the first United States patent statute became law on April 10, 1790. The celebration had been suggested on time, but there was no organization to arrange it. Finally, after the celebration did not occur on time, an ad hoc

[12]See [Edi86], [Lad02, p. 753], [Ita83], and [Tak36].

[13]See *Washington Post* 8/16/1891 p 16 and [dW58]

[14]See [Edi87].

committee, chaired by patent attorney Robert Washington Fenwick, moved the plans off center, and momentum took over .... See Prologue—The Centennial Celebration.

Epilogue: Fast Forward to the Early Third Century

The importance of the Model Room continued to decline. Richard C. Gill, chief of the Model Room for many years, was the son of John Gill, the inventor of a revolving gun, mentioned in Chapter 15. In 1901, Richard Gill wrote that the importance of the models was fast declining, as shown by the decline in the interest in them. Prior to the fire of 1877, it took 15 model attendants to allow inspection of the models, but by 1901 the work was done by three attendants. Strangers visiting the office might ask to see the model of a famous invention, but they were also likely to ask for "my father's" or "my grandfather's" model.[1] Soon afterwards, the models were moved to rented spaces in the Union Building, where they could still be visited with some difficulty, but seldom were. The Model Rooms were reconfigured as offices. By 1908, the models were packed into boxes and stored elsewhere, and the rented spaces in the Union Building were given up. Finally, in 1925, the important models were given to various museums. A very few models were kept by the Patent Office for exhibit there. The rest were sold at public auction, the first third of them selling for $1,575. Most of the models went to a purchaser who used them for a short period in a Patent Model Museum in New York City. They were soon in storage in barns in Garrison, New York. Only in recent years has some interest been taken in these models.[2]

On January 10, 1912, Commissioner Edward B. Moore testified before the House Committee on Patents that until very recently the Patent Office had required working models to prove operability of perpetual motion machines and man-carrying heavier-than-air flying machines. When the Wright brothers

[1] See R.C. Gill to E.V. Shepard 8/1/1901 National Archives Suitland Accession No. 55-A-584 Record Group 241 Box 133.

[2] See [Bre26].

demonstrated in 1903 that they could fly, the model requirement for flying machines was dropped but not for perpetual motion machines.[3]

Until 1897, anyone could practice before the Patent Office who had not been disbarred and forbidden to do so by the Patent Office. It was not unusual for patent solicitors to begin practice at the age of 16 to 18. The only requirement was to find a trusting client. For years, the Patent Office had wanted to control who practiced before the office. Commissioner Butterworth issued an order on August 6, 1897, requiring registration of patent practitioners. The first registrations occurred on August 3, 1897, three days before the notice was actually issued. The first registrations were under a grandfather clause. One could be qualified for registration by having prosecuted a patent application before the Patent Office within the past five years.

Registration number 1 was assigned to Gales Pritchard Moore (1873–1939) of Washington, D.C. Registration number 2 was assigned to Charles W. Gardner of Newaygo, Michigan. The roster of registered patent attorneys as of January 1, 1899, included registration numbers up to about 2,550.

Gales is an old Washington name. Joseph Gales Sr., and Joseph Gales Jr., were successively the publishers of the old National Intelligencer newspaper, published from about 1800 to the 1850s in Washington. Joseph Gales Jr., was once the mayor of Washington. Captain William W. Moore, a militia captain, had been a printing foreman at Mr. Gales' newspaper and was later the secretary and treasurer of the metropolitan railroad—the streetcar company. He named his son Joseph Gales Moore after his employer. Joseph Gales Moore was a cashier at the Metropolitan Bank of Washington; he married Kate Carroll Pritchard, and their son was Gales Pritchard Moore, born 1873.

Gales P. Moore received an LL.B. from the George Washington University in 1894, an LL.M. in 1895, and an MPL in 1896. He was first employed as a clerk in the office of Charles L. Sturtevant, a Washington patent lawyer. In 1899, he began practice in Washington as a sole patent practitioner, and in 1901 moved to St. Louis to work for Baker and Cornwall, patent attorneys. In 1904, he moved to Bristol, Connecticut, to become patent counsel for the New Departure Manufacturing Company, manufacturer of bicycle coaster brakes. In 1922, New Departure was acquired by General Motors, and Mr. Moore became Chief Patent Counsel of the Eastern Division of General Motors, remaining until his death in Bristol in 1939. He was always proud to be the registered patent attorney with registration number 1.[4]

[3]See Congressional Information Services microfiche series House Committee on Patents 62-C.

[4]See Obituary *The Bristol Press* 10/4/1939; 1899 *Roster of Registered Patent Attorneys*; various Washington City Directories.

A Department of Commerce and Labor was established by the Act of February 14, 1903, and split into a Department of Commerce and a Department of Labor in 1913. The 1903 act had authorized the President to place any scientific office from other departments including the Department of Interior into the Department of Commerce and Labor. By a 1925 Executive Order, the President transferred the Patent Office from Interior to the Department of Commerce as of April 1, 1925. The offices of the Department of the Interior, other than the Patent Office, had moved out of the Patent Office Building to a new departmental building in 1917. For a brief period of time, the Patent Office was the sole occupant of the Patent Office Building.

From its origin until 1922, the Patent Office paid its own way and generated a surplus. There were a few individual years in the red, but overall, the Patent Office had a surplus. This ended about 1922, followed by at least 70 years of operation with a deficit.

In 1932, after 92 years in the Patent Office Building, the Patent Office moved into the newly completed Herbert Hoover Building, the home of the Department of Commerce, between Constitution Avenue and E Street near Pennsylvania Avenue on two sides and 14th and 15th Streets on the other two sides. When the building was begun, it was overlooked that, as late as the Civil War, the area had been the site of Tiber Creek, flowing through Washington on its way to the Potomac. It was encased in an aqueduct and buried for years, but when construction began on the new building, it burst forth into the newly created excavation. A deep-sea diver was called in from Philadelphia to again encase Tiber Creek before work could be continued. To this day, the Tiber Creek aqueduct carries hidden and mostly forgotten Tiber Creek through the basements of the Herbert Hoover Building.

Patent Office End of Herbert Hoover Building

The entire Commerce Building, rising seven floors and covering eight acres of land, provides over a million square feet of floor area. The north end of this building, overlooking Pennsylvania Avenue, was specially designed for the Patent Office. In the stone over its north end of its west side was carved Abraham Lincoln's remark: "The Patent System Added the Fuel of Interest to the Fire of Genius." Electricity was fully provided, to be used for lighting, fans, elevators, vacuum cleaners, cook stoves in the cafeteria, every purpose for which electricity could be used in what was then the largest and most modern government building ever built.[5] While the electrically operated elevators required the Commerce Department to hire numerous elevator operators, electricity also eliminated the previously required clock-winders who went through offices daily, winding and setting clocks.

Chester Floyd Carlson (1906–1968) was a patent attorney who needed a simple way to copy patent drawings and papers, so he invented xerography in 1938 and impoverished himself while developing it. Eventually his invention was adopted by Haloid Corporation, which developed into Xerox Corporation. Carlson's invention resulted in the final displacement of the public typists who had replaced the Patent Office copyists of the previous century.[6]

Ten years after the Patent Office was moved into the Commerce Building, it was moved out of the building and out of the city. On December 27, 1941, twenty days after the attack on Pearl Harbor, Conway P. Coe, Commissioner since 1933, called together the employees of the Patent Office and announced that because of the sudden need for space in Washington to coordinate the prosecution of World War II, most of the divisions of the Patent Office were to be moved to Richmond, Virginia, to occupy the old Export Leaf Tobacco Company warehouse at 900 North Lombardy Street.[7]

This was met with vehement protest. The administration may not have realized that the location of the Patent Office concerned not only a group of presumably docile government employees, but also a multitude of patent attorneys who arranged their lives in accordance with the location of the Patent Office. Employees who were to be transferred complained that the available housing in Richmond was of eighteenth-century vintage. The tobacco warehouse was cleaned, remodeled and painted. The first air-conditioning ever available in the Patent Office was available there, having formerly been used to condition the tobacco stored there. It was very loud, but it only ran four minutes each hour. When the outside of the building was painted white, local residents complained that the repainted building made Richmond an exceedingly conspicuous tar-

[5]See [Edi32].

[6]See [Des71] and [Hey76].

[7]See [E.M42].

get. Officials pointed out that the State Capitol was equally prominent and would give equally as much help to the enemy in locating Richmond.[8]

By 1944, the employees of the Patent Office were, for the most part, settled into Richmond. In 1943, employees had started their own orchestra, and by the spring of 1944 the Patent Office Concert Orchestra was presenting concerts in public halls in Richmond. All members of the orchestra were either employees of the Patent Office or their wives.[9] Employees consistently exceeded their goals for Community Fund contributions to the Richmond charity by a third or a half. In October of 1944, rumors began to circulate about the return to Washington. But return was not to be immediate. In January 1945, it was announced that the interference division and the docket division would be returned to Washington within a month. By July, when plans were being considered for eventual return, 250 of the 900 Patent Office employees had been hired in Richmond and were not likely to go to Washington. Commissioner Coe left office in Sept 1945 and was immediately replaced by Commissioner Casper W. Ooms. In October, Commissioner Ooms announced that some of the divisions might return by the end of the year. The last three Patent Office employees left Richmond, Virginia, in October 1946.[10]

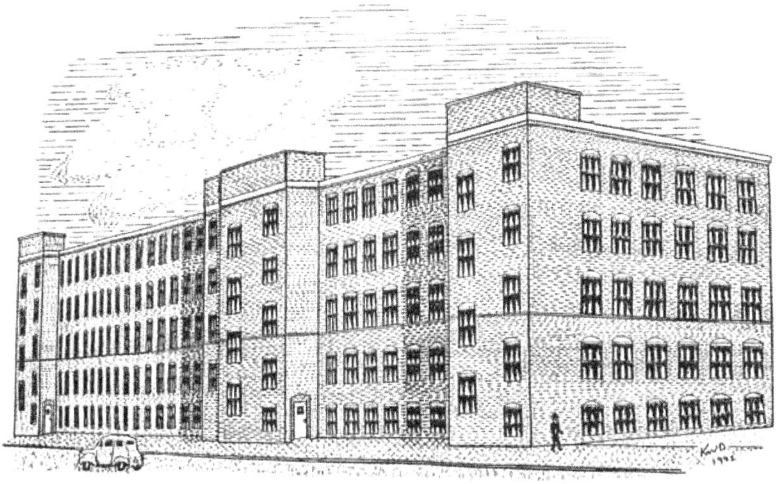

Patent Office in Richmond Tobacco Warehouse

[8]See *Richmond News-Leader* 12/30/1941, 1/6/1942, 2/20/1942, and 3/18/1942.
[9]See *Richmond News-Leader* 10/14/1944.
[10]See *Richmond News-Leader* 10//14/1944, 7/10/1945, 10/17/1945, and 10/22/1946.

During the Richmond years, a truck made daily deliveries between Washington and Richmond. Some Patent Office employees made a daily trip by train or car pool between the cities. Efficiency dropped 25 percent during the exile. There were still 453 Patent Office employees working in Richmond when the issue was forced, in September 1946, by moving them temporarily into Temporary Building No. 7 at the north end of Washington National Airport at Gravelly Point, formerly used by the Army Air Corps.[11] Eventually all units were returned to the Commerce Building.

The Patent Office in the Commerce Building was overcrowded and uncomfortable. It was mostly not air-conditioned and could be very hot in summer. The Commerce Building by the 1960s was seriously in need of massive renovation. A search was made for a convenient place for the Patent Office. Many distant communities wanted the office in their cities, but it was decided to keep it in the Washington area. Charles E. Smith Company was developing Crystal City, along the Potomac in Arlington, Virginia, just inland from National Airport. This was the chosen location. It was three miles from the White House and from the Commerce Building. For attorneys from out of town, it was closer to the new hub of transportation, National Airport, than either the Commerce Building or the old Patent Office were to the old hub of transportation, Union Station. It was the first long-term Patent Office home in rented quarters.

Initially, the Patent Office was to occupy two eleven-floor buildings called Crystal Plaza 3 and 4, together with a lower building connecting these two called Crystal Plaza 3-4. A third building, Crystal Plaza 2, was soon added to the Patent Office. The first Patent Office unit to move was the Electrical Examining Operation, which began moving in April 1967. The Public Search Room began its move in December 1967. The total move was completed in about two years.[12] Afterwards, it spread to many other nearby buildings.

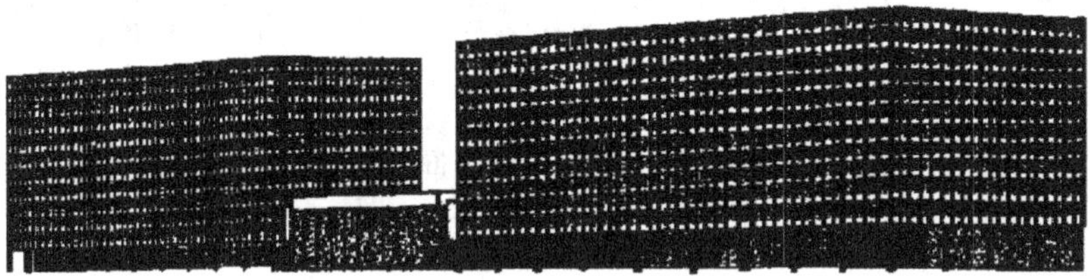

Patent Office in Crystal Plaza 3 and 4, Arlington, VA

[11]See *Washington Star* 10/23/1945 and 9/16/1946, [Kal46].
[12]See [Mer68].

In 1975, the Patent Office was renamed the Patent and Trademark Office (usually abbreviated PTO) to recognize the importance of the trademark side of the office.[13]

By 2005, the PTO had moved into a newly constructed multiple-building[14] complex of its own at Randolph Square in Alexandria, Virginia, leaving Crystal City behind.[15] It has opened satellite offices in Detroit, Denver, San Jose and Dallas. A decade after moving to Alexandria, the PTO does not have sufficient room in its Randolph Square complex to provide office space for all of its examiners, and many by necessity are urged to telecommute by working from computers in their own homes. This is called teleworking. When a teleworker has to come in on occasion to work in the office, it is called hoteling, because office space will be assigned to the teleworker for the occasion. In 2014 about 75 percent of the total PTO workforce worked from home at least one day a week. This means that, if teleworkers work from home only one day per week, the PTO has at least fifteen percent more workers than it has offices to put them in. And many employees telework more days per week than that.[16]

According to C. Northcote Parkinson,[17] writing of corporate headquarters and other buildings, "A perfection of planned layout is achieved only by institutions on the point of collapse." In other words, when you can design a building which is perfect for your intended purposes, your purposes have been frozen, and decline has set in. If so, the PTO can rest assured that its purposes are still expanding, and that everything that can be invented has not yet been invented. The PTO outgrew the building designed especially for it less than ten years after it was designed.

In the early days, all patent applications were handwritten. A few years back, the PTO added a requirement that all application papers must be typewritten or mechanically printed. This requirement went almost unnoticed, because almost nobody had been filing handwritten applications for many years. More recently, a requirement has been added to file applications electronically

[13]See United States Patent and Trademark Office, General Information Concerning Patents, 2014, published by the PTO on the internet.

[14]For the curious, these are Jefferson Building, Knox Building, Madison Building East, Madison Building West, Randolph Building, Remsen Building, Carlyle Townhouse, Elizabeth Townhouse, and East and West Parking Garages.

[15]See United States Patent and Trademark Office, Performance and Accountability Report for Fiscal Year 2005. published by the PTO on the internet.

[16]See United States Patent and Trademark Office, Performance and Accountability Report for Fiscal Year 2014, published by the PTO on the internet.

[17]See [Par57].

or to pay a fee to convert typewritten papers to an electronic format within the PTO. Examiners no longer work routinely with paper.[18]

The PTO changed its name formally to the United States Patent and Trademark Office in 2000.[19] Perhaps this was needed as a show of international cooperation, but it is still abbreviatable as PTO. One reason for changing one's name can be to hide one's past. The PTO has absolutely no reason to do that and every reason to proclaim it before the world.

Beginning in 1998, the head of the Patent and Trademark Office was given the title "Under Secretary of Commerce for Intellectual Property and Commissioner of Patents and Trademarks." Then, on January 17, 2001, the head of the Patent and Trademark Office became the "Under Secretary of Commerce for Intellectual Property and Director of the United States Patent and Trademark Office." Under this new title, there are a "Commissioner of Patents" and a "Commissioner of Trademarks" who report to the Director. So there is still a Commissioner of Patents down in the hierarchy somewhere. And, in this modern age, the revolution started in the Patent Office by Clara Barton has been completed; the current Director of the United States Patent and Trademark Office is female—Michelle K. Lee

A new patent statute[20] recently took effect which does away with the interferences that the Patent Office or the PTO used to declare between two or more patent applications (or between an application and an issued patent) claiming the same thing. An interference proceeding would determine which applicant had priority (and would get a patent). The new patent statute also did away with a previously existing procedure called an inter partes reexamination. Since its earliest days, the Patent Office has operated to give a patent to the first diligent inventor to invent something. Under the new statute, with a few exceptions, the patent for an invention goes to the first inventor to file an application for patent on that invention. It is now a "race to the Patent Office." This was part of an effort to make U.S. patent laws compatible with those of the rest of the world. For years, only the United States, Canada and the Philippines had interference procedures in their patent law. When first Canada and then the Philippines gave them up, the United States could not be far behind. The new statute also offers the opportunity for third parties to oppose the continued validity of a patent after it has been granted. Prior art available to prevent patenting an invention has been expanded to include everything "available to the public before the effective filing date of the claimed

[18]See United States Patent and Trademark Office, General Information Concerning Patents, 2014, published by the PTO on the internet.

[19]See United States Patent and Trademark Office, General Information Concerning Patents, 2014, published by the PTO on the internet.

[20]See Leahy-Smith America Invents Act, Public Law 112-29, effective September 16, 2011.

invention." Prior art now includes items sold in an artisan's shop in Italy or described in a blog on the internet.

The American inventor is an extraordinary creature. He invents in any field he is interested in, which includes every field that anyone else is interested in, and many that he shares with no one else—yet. Any time you think you understand him and can predict what he might do, you may find to your surprise that you really do not understand her at all.

The Franklin Institute is still active, and its Journal is still published regularly, although no longer directly by the Franklin Institute. Scientific American is now published as a preeminent popular scientific journal, independent from its former patent connections. When the Patent Office decided that patent law firms should no longer be allowed to advertise, the patent law firm of Munn & Co. disassociated itself from its former journal and continued the practice of patent law until shortly after the death in 1958 of Orson D. Munn, the grandson of the founder. It no longer exists. The National Museum continues, now part of the Smithsonian Institution. The Department of Agriculture continues, having long since outstripped in size its parent, the Patent Office. When the Patent Office began the patenting of certain plants, the Agriculture Department once again became loosely associated with the Patent Office as to those patents. In mid-1994, the law firm of Mason, Fenwick and Lawrence, in its fourteenth decade, merged with Popham, Haik, Schnobrich & Kaufman, Ltd. The last attorney named Fenwick had retired from the firm, leaving one Fenwick descendent in the combined firm.

The old Patent Office building, from which the Patent Office moved in 1932, has been converted into two art galleries. The south end, occupied by the Smithsonian's National Portrait Gallery, is particularly interesting for its third-floor exhibits and for the old iron galleries surrounding its library.

Annie Ellsworth's direct descendant in the all-female line, Miss Jennie Jackson of Medford, Oregon, in the early 1990s married Brian Morse, Samuel Morse's only direct descendant in the all-male line in Brian's generation. They met in the eighth grade and were immediately smitten, but did not know of the connection between their ancestors until her grandmother told them about it after they had announced their engagement as juniors in college. Annie Ellsworth's crush was still working 150 years later.

The Patent Office, under its new name, the Patent and Trademark Office, did business from its location across the Potomac River in Arlington, Virginia, until it moved to a new campus in Alexandria, Virginia in 2005. It has now issued well in excess of nine million patents. When the Patent Office burned in 1836, 10,000 patents—all of the patents that had been issued in 46 years—were burned. The Patent and Trademark Office of 2016 issues 10,000 patents on average in about a week and a half. Almost half of its patents were issued since the first edition of this book.

Oil paintings of William Thornton and of Robert Fulton were hung side by side in the Crystal City office of the Commissioner of Patents and Trademarks, although at the peak of their feud the gentlemen themselves would not have stood so close to each other.

The Patent Office pony was successively assisted by the horse-drawn streetcar, the locomotive, the telegraph, the telephone, the automobile, and the fax machine, each soon after its first appearance in the Patent Office as an application for patent. It was also assisted by the internet, but that had no clear single inventor, despite suggestions to the contrary by Al Gore. But in spirit at least, the Patent Office pony is still ridden in Alexandria every workday.

Oh, and *John Smith*, our Everyman the Inventor, born in 1776, 1777, 1778, and every succeeding year, was born again this year. Even if we stifle him, we will not be able to prevent his flights of imagination, but if we convince him that his dreams are impossible, they will be. If we encourage him, he will send us to the stars, conquer contagious diseases one by one, invent machines which will be proud of their inventors, provide limitless safe sources of energy, harness physical laws which are yet to be discovered, and create things that we cannot now dream of and later will not be able to conceive of doing without.

Here ends the guided tour. Your guide, who once studied drafting but never studied real art, apologizes for subjecting you to his amateur artistic efforts throughout the book. In mitigation, it could be said that no other source was available for the pictures thus produced. Where drawings by skilled artists were available, they were used in the book.

Appendix A

Some Civil War Era Patents

List of Confederate Patents

The following is a list of all known patents issued by the Confederate Patent Office. Each entry in the list consists of the patent number, followed by the date of issue, the inventor's name, the inventor's address, and finally the title of the patent. Your guide had been unable to discover if any patents were issued after 01/01/1865. Recently a researcher in the records held by the Museum of the Confederacy[1] found assignment records to indicate that patents 267–274 were issued in 1865.

An asterisk (*) precedes an alternate version of text found in the CSA Patent Office Reports.[2]

1. 08/01/1861 J.H. Van Houten of Savannah, GA for Breech Loading Gun

2. 08/01/1861 Charles E. Brown of Staunton, VA for Stirrups

3. 08/05/1861 George B. Sloat of Richmond, VA for Sewing Machine Stand, Design

4. 08/05/1861 George B. Sloat of Richmond, VA for Work Box for Sewing Machine, Design

[1] See [Kni11].

[2] See [Rho65], four annual reports.

5. 08/06/1861 R.W. Habersham of Beaufort, SC for Sabre Lance
6. 08/08/1861 James W. Frazier of Frazierville, SC for Wheels for Steamboats
7. 08/07/1861 John M. Walden of Fort Valley, GA for Switch for Railroads
8. 08/10/1861 Phidello Hall of Springfield, TX for Automatic Breech Loading Gun
9. 08/12/1861 Thomas W. Cofer of Portsmouth, VA for Revolving Pistol
10. 08/13/1861 John D. Evans of Pleasant Hill, GA for Plough
11. 08/15/1861 Armand Preot of Gravel Hill, VA for Lance or Pike, Attaching to Guns
12. 08/17/1861 Frederick J. Gardner of Newbern, NC for Cartridges
13. 08/23/1861 J.R. (*P.R.) Clements of Eufala, AL for Water Wheels for Mills
14. 08/24/1861 Victor Armant of New Orleans, LA for Apparatus for Clarifying Sugar-Cane Juice
15. 08/26/1861 J.L. Jones of Tallyhoe, NC for Carriage Wheels
16. 08/27/1861 J.D. Stewart (*Steward) of Jackson, MS for Sash Fastener
17. 08/29/1861 W.J. Cheshire of Colquitt, GA for Plough
18. 09/02/1861 E. Boyle, T. Gamble and E. Macfee of Richmond, VA for Sword Bayonet Attaching to Guns
19. 09/03/1861 James H. Harkut (*Carcut) of Natchez, MS for Manufacture of Cannon
20. 09/07/1861 J.S. Boothby of Savannah, GA for Tanning
21. 09/14/1861 James P. Rankin of Marion, NC for Breech Loading Gun
22. 09/16/1861 Isaac Beirfield of Newberry C.H., SC for Tanning
23. 09/19/1861 Ed. Gotthiell of New Orleans, LA for Percussion Fuses
24. 09/27/1861 E.T. Ligon of Demopolis, AL for Breech Loading Firearms

25. 09/28/1861 Benjamin Winter of Buckingham C.H., VA for Winnowing Machine

26. 10/01/1861 John R. Spilman (*Spillman) of Warrenton, VA for Cartridge Paper

27. 10/02/1861 John R. Spilman (*Spillman) of Warrenton, VA for Device for Making Cartridges

28. 10/02/1861 James S. Allums of Cussetta, GA for Cotton Presses

29. 10/04/1861 James Lynch of Petersburg, VA for Cannon

30. 10/07/1861 Arthur E. Hall of South Quay, VA for Camp Bedstead

31. 10/10/1861 E.J. Park of Memphis, TN for Tannin

32. 10/08/1861 E.N. Davis of Holly Springs, MS for Cotton Ties

33. 10/11/1861 Robert C. Bernard of Rocky Mount, VA for Gate Fasteners

34. 10/15/1861 Joseph Thomas of Batesville, AR for Bullet Moulds

35. 10/15/1861 Daniel Oswalt of Cubahatachie, AL for Breech Loading Cannon

36. 09/01/1861 (*09/10/1861) Alfred G. Hearn of Arkadelphia, AR for Machine for Measuring Distances

37. 10/24/1861 Hannibal S. Blood of New Orleans, LA for Switches and Turntables for Horse Railroads

38. 10/29/1861 R.C. Howe of Richmond, VA for Camp Chest and Bedstead Combined

39. 10/30/1861 Henry C. Goodrich of Augusta, GA for Camp Cot

40. 10/02/1861 John P. Gorman of Charlestown, VA for Cartridge Boxes

41. 11/07/1861 William W. Rickenbacker of Beaufort, SC for Camp Chest and Bed Combined

42. 11/07/1861 Cornelius Cox of Cahaba, AL for Cotton Cleaner

43. 11/11/1861 T.W. Moore of Person Co., NC for Plough

44. 11/11/1861 M.A. McLeod of Thibodaux, LA for Apparatus for Clarifying Sugar-Cane Juice

45. 11/20/1861 George W. Peabody of Columbus, GA for Waterproof Composition

46. 11/21/1861 R.W. Biggs of Jacksonville, FL for Plough

47. 11/22/1861 Andrew Day of Woodville, MS for Plough

48. 11/26/1861 John N. Gradick of Fayetteville, AL for Washing Machine

49. 07/23/1861 L.B. Woolfork of Goodlettsville, TN for Steam Plough

50. 11/21/1861 John Schley of Savannah, GA for Horse Power

51. 11/27/1861 Columbus A. Rose of Columbia, AL for Hoes

52. 11/27/1861 (*11/07/1861) B.B. & W.H Stephens of Dallas, LA for Corn and Cottonstalk Cutter and Puller

53. 12/06/1861 Louise Grady of Norfolk, VA for Washing Machine

54. 10/07/1861 John M. White of Citrouille, AL for Breech Loading Guns

55. 10/11/1861 Marcus A. Tarleton of New Orleans, LA for Cotton Ties

56. 10/14/1861 Henry Domler of Wilmington, NC for Military Cap

57. 10/31/1861 W.B. Martin of Fayetteville, TN for Sewing Machine

58. 01/02/1862 (*01/21/1861) Carl Laquequist of Macon, GA for Breech Loading Guns

59. 01/04/1862 J.S. Peete of Tipton, TN for Cotton Ties

60. 01/07/1862 Jacob B. and William L. Platt of Augusta, GA for Camp Cots

61. 01/08/1862 J. Nichols and J. Bennett, assignors to F. Smith of Memphis, TN for Submarine Battery

62. 01/09/1862 R.P. Moore and N. Thompson of Box Springs, GA for Mosquito and Fly Brush Machine

63. 01/09/1862 A.P. (*A.R.) Routt of Somerset, VA for Seed Planter

64. 01/09/1862 John B. Wand of Memphis, TN for Waterproof Cloth

65. 01/20/1862 G.W. Dolbey of Carrollton, MS for Bellows

66. 01/29/1862 John Wells of Holly Springs, MS for Cotton Press
67. 01/31/1862 James Peeler of Bartow, GA for Ploughs
68. 02/01/1862 M.G. (*G.M.) Rhodes and A. Bingham of Talladega, AL for Wooden Bottomed Shoes
69. 02/03/1862 G.W. Dolbey of Carrollton (*Carrolton), MS for Machine for Shrinking Wagon Tires
70. 01/07/1862 R. Archer of Richmond, VA for Percussion Fuse
71. 02/05/1862 J.F. Finger of Marion C.H., SC (*VA) for Portable Saw Mill
72. 02/05/1862 John Dove of Lauderdale (*Lunderdale) Sta., MS for Cotton Cleaning Machine
73. 02/05/1862 A. Knapp of Mobile, AL for Mosquito and Fly Repelling Machine
74. 02/15/1862 William Spillman of Mobile, AL for Machine for Making Lead Pipes and Bars
75. 02/21/1862 Benjamin Dennis of Scottsdale, VA for Evaporator
76. 02/25/1862 John D. Love of Wilmington, NC for Combination Bedstead
77. 02/28/1862 William Conner of Natchez, MS for Cotton Tie
78. 02/28/1862 Abraham Hagar of New Orleans, LA for Bagasse Furnace
79. 03/11/1862 C.V. Littlepage of Austin, TX for Bullet Machine
80. 03/07/1862 B.C. Hattox of Little Rock, AR for Tanning
81. 03/14/1862 T.S. (*T.C.) Copes of New Orleans, LA for Cotton Tie
82. 03/14/1862 Thomas Dale of Russellville, KY for Device for Replacing Locomotives on Railroad Cars
83. 03/21/1862 P.R. (*R.P.) Clements of Eufala, AL for Looms
84. 03/21/1862 Charles E. Stuart, J.C. Owing and J.H.C. Taylor of Alexandria, VA for Instrument for Sighting Cannon
85. 03/24/1862 Gardner Smith of New Orleans, LA for Rail for Railroad Cars (*Curves)

86. 03/25/1862 John M. Morehead of Greensboro, NC for Heating Apparatus
87. 03/29/1862 John E. Pattison of Houma, LA for Apparatus for Clarifying Cane Juice
88. 04/14/1862 Augustus McBurthe of Richmond, VA for Mode of Manufacturing Scabbards
89. 05/05/1862 William Hicks of Henderson, TX for Pumps
90. 05/10/1862 E.B. Stephens, assignor to J.A. Van Lew of Charleston, SC for Portable Flat Boat
91. 05/10/1862 J.W. Howlett of Greensboro, NC for Breech Loading Fire Arms
92. 05/01/1862 C.A. McEvoy of Richmond, VA for Fuse
93. 05/23/1862 C.A. McEvoy of Richmond, VA for Fuse
94. 05/03/1862 W.P. Wyley (*W.B. Wylly) and A.M. Barbee of Savannah, GA for Railroad Car Oil Box
95. 06/15/1862 James I. Roberts of White Springs, FL for Plough
96. 05/31/1862 J.G. (*G.J.) Peterson of Marion, NC for Fire Arms
97. 07/10/1862 Robert Buillock of South Mills, NC for Plough
98. 07/22/1862 John Cowdon (*Cawdon) of New Orleans, LA for Vessel of War
99. 07/22/1862 J.A. (*S.A.) LeToudal of Mobile, AL for Tent
100. 07/29/1862 John M. Brooke of Richmond, VA for Ship of War
101. 07/29/1862 James E. Watson of Petersburg, VA for Combined Knapsack and Tent
102. 07/31/1862 John Commins of Charleston, SC for Tanning
103. 08/01/1862 John Commins of Charleston, SC for Tanning Vat
104. 08/06/1862 William S. Morris, of Lynchburg, VA for Composition for Galvanic Batteries
105. 08/16/1862 J.M. Jackson of Columbus, MS for Chain Link

106. 09/25/1862 Lucien Hopson of Lampassas, TX for Projectile

107. 09/26/1862 John M. Brooke of Richmond, VA for Reversible Hook

108. 09/27/1862 George Henry of Columbus, MS for Breech Loading Fire Arms

109. 09/29/1862 J.A. LeToudal of Mobile, AL for Instrument for Leveling Cannon

110. 10/01/1862 Augustus McBurthe of Richmond, VA for Waterproof Cloth

111. 10/13/1862 (*10/03/1862) A.J.M.T.O. Clanton of Panola, MS for Breech Loading Firearms

112. 10/13/1862 William Spillman of Prattville, AL for Bullet Machine

113. 10/11/1862 William S. Winfield of Springfield, TN for Projectile

114. 10/15/1862 (*10/14/1862) James C. Patton of Petersburg, VA for Submarine Battery

115. 10/17/1862 John Hollander of Sutherland Springs, TX for Composition for Toilet Soap

116. 10/17/1862 John Hollander of Sutherland Springs, TX for Compound for Expelling Cattle Worms

117. 10/17/1862 John Hollander of Sutherland Springs, TX for Composition for Washing Soap

118. 10/24/1862 James H. Douthatt of Montgomery, VA for Machine for Making Shoe Pegs

119. 11/08/1862 (*11/05/1862) G.H. Whitescarver of Richmond, VA for Protractor

120. 11/08/1862 (*11/05/1862) G.H. Whitescarver of Richmond, VA for Surveying Instruments

121. 11/08/1862 (*11/05/1862) D.R. Williams of Richmond, VA for Breech Loading Cannon

122. 11/10/1862 Charles Mahon of Macon, GA for Postage Stamp Case

123. 11/15/1862 C.A. McEvoy of Richmond, VA for Fuse

124. 11/13/1862 J.E. Jenkins of Pelahatchie, MS for Plough

125. 11/20/1862 M. Bridges of Memphis, TN for Breech Loading Cannon

126. 12/?/1862 Wiley Futrell of Oxford, GA for Rotary Plough

127. 12/08/1862 J.G. Wire of New Orleans, LA for Machine for Operating Submarine Batteries

128. 12/10/1862 James A. Cameron of Canton, MS for Varnish

129. 12/11/1862 Nathaniel Nuckolls of Russell Co., AL for Army Canteen

130. 12/30/1862 E.S. Collins of Aspinwall. VA for Wooden Soled Shoe

131. 12/19/1862 John Henson of Fork Union, VA for Hay Press

132. 12/30/1862 Joseph A. Yates of Charleston, SC for Machinery for Traversing Guns

133. 01/03/1863 Alonzo C. Chinn of Mobile, AL for Projectile

134. 01/07/1863 Henry T. Beard of Emery, MS for Composition for Finishing Leather

135. 01/07/1863 C.A. McEvoy of Richmond, VA for Fuze

136. 01/15/1863 George H. Lehner of Richmond, VA for Clamps for Lathes

137. 01/15/1863 George H. Lehner of Richmond, VA for Turning Lathe

138. 01/15/1863 George H. Lehner of Richmond, VA for Lathe Tool

139. 01/15/1863 Henry G. Street of Warren Co., MS for Cotton Scraper and Plough

140. 01/27/1863 William S. Watkins of Houston, TX for Wind Mill

141. 01/15/1863 Lawson G. Peel of Preston, GA for Plough

142. 01/20/1863 G.S. Newman, assignor to A.P. Routt (*Rolett), of Somerset, VA for Plough

143. 01/26/1863 R.H. Barrett of Murfreesboro, NC for Projectile

144. 01/26/1863 Thomas Ashcroft of Talledega, AL for Machine for Making Combs

145. 02/02/1863 Charles H. Winston of Richmond, VA for Method of Constructing Sulphuric Acid Chamber

146. 02/02/1863 B.H. Washington of Hannibal, MO for Lamp Burner

147. 02/14/1863 Henry F. Cromwell of Cynthiana, KY for Gang Plough

148. 02/14/1863 J.H. Tarply of Greensboro, NC for Breech Loading Fire Arms

149. 02/18/1863 D.W. Hughes of Don Juan, AZ for Breech Plug

150. 03/10/1863 J.A. Park of Sequin, TX for Tanning

151. 03/10/1863 H. Leyden (*A. Legden) of Atlanta, GA for Revolving Fire Arms

152. 03/13/1863 E.S. Cook of Laurel Grove, VA for Plough

153. 03/13/1863 Felix Walker of New Orleans, LA for Mode of Packing Cotton

154. 03/20/1863 N.T. Read of Danville, VA for Breech Loading Fire Arms

155. 03/23/1863 James E. Hanger of Staunton, VA for Artificial Limb

156. 03/26/1863 George H. Lehner of Richmond, VA for Sewing Machine Shuttle

157. 03/31/1863 W.D. Mason of Jarratt's Depot, VA for Water Wheel

158. 04/06/1863 George H. Lenher of Richmond, VA for Lathe

159. 04/16/1863 O.L. Bayley (*Bailey) of Dallas, TX for Gang Ploughs

160. 04/15/1863 Thomas H. Bacon of Hannibal, MO for Projectile

161. 04/15/1863 Ellison Yerby of Richmond, VA for Machine for Filling Cartridges

162. 04/18/1863 John J. Daly of New Orleans, LA for Instrument for Measuring Distances

163. 04/18/1863 C.W. Alexander of Moorfield, VA for Breech Loading Fire Arms

164. 04/20/1863 John B. Norris of Clark Co., VA for Fire Tongs

165. 05/01/1863 T.L. Jones of Natchez, MS for Cotton Tie

166. 05/01/1863 T.L. Jones of Natchez, MS for Cotton Tie

167. 05/01/1863 (*05/21/1863) Robert Creuzbaur of Travis Co., TX for Wood Soled Shoe

168. 05/04/1863 F.D. Lee of Charleston, SC for Spar Torpedo

169. 05/06/1863 R.S. Dunning of Atlanta, GA for Machine for Cutting Irregular Forms

170. 05/09/1863 George H. Lehner of Richmond, VA for Sewing Machine Shuttle

171. 05/20/1863 (*05/21/1863) Sylvester L. Burford of Lynchburg, VA for Wooden Shoe Sole

172. 05/25/1863 Alfred L. Bowen of Winchester, VA for Button Machine

173. 05/25/1863 F.F. Taber of Atlanta, GA for Process of Producing Stearine from Tallow

174. 05/25/1863 William Le Roy Browne of Ivy Depot, VA for Fuze

175. 05/26/1863 Charles C. Bier of New Orleans, LA for Cotton Tie

176. 06/05/1863 Isidor P. Girardy of Augusta, GA for Fuze

177. 06/05/1863 A.C. Elder of Winnsboro, SC for Churn

178. 06/10/1863 Asa George of Charlotte, NC for Revolving Fire Arm

179. 06/18/1863 S.A. Hinkel of Lewisburg, VA for Tanning

180. 06/22/1863 William H. Wilson of Staunton, VA for Match Cutting Machine

181. 07/10/1863 Benjamin M. Cook of Kinston, NC for Explosive Apparatus

182. 07/15/1863 Albert Bloomer, assignor to W.N. Smith of Richmond, VA for Fuze Gauge and Cutter

183. 07/15/1863 John Stillwell of Griffin, GA for Churn

184. 07/22/1863 Charles Knowlton of Talboton, GA for Machine for Cutting Shoe Pegs

185. 07/22/1863 Phillip E. Love of Augusta, GA for Projectile

186. 07/25/1863 Miles Waterhouse and William Wheelan (*Whealan) of Charlotte, NC for Varnish

187. 08/05/1863 A. Barbarin of New Orleans, LA for Torpedo

188. 08/05/1863 (*08/10/1863) Homer Holland of Charlotte, NC for Process of Producing Copperas

189. 08/10/1863 Hughes, Pendergrass and Snow of Monroe, GA for Loom

190. 08/10/1863 Martin B. Tyler of Richmond, VA for Hominy Mill

191. 08/22/1863 H. Conner of Mobile, AL for Projectile

192. 08/22/1863 H. Conner of Mobile, AL for Breech Loading Cannon

193. 08/25/1863 A.F. Volck of Nuremburg, Bavaria for Bust of Lt. Gen. T.J. Jackson, Design

194. 08/25/1863 R.J. White and George Lehner of Richmond, VA for Bullet Machine

195. 08/28/1863 Robert Creuzbaur (*Creugbaur) of Austin, TX for Wooden Sole Shoe

196. 08/28/1863 Robert Creuzbaur (*Creugbaur) of Austin, TX for Half Wooden Sole Shoe

197. 08/31/1863 Robert Creuzbaur (*Creugbaur) of Austin, TX for Torpedo

198. 09/21/1863 Z. McDaniel of Glasgow, KY for Torpedo

199. 09/10/1863 Thomas Morse of Richmond, VA for Breech Loading Firearm

200. 09/15/1863 E.C. Singer of Port Lavaca, TX for Torpedo

201. 08/18/1863 James E. Hanger of Staunton, VA for Artificial Leg

202. 09/21/1863 Z. McDaniel of Glasgow, KY for Torpedo

203. 09/24/1863 B.H. Washington of Hannibal, MO for Machine for Casting Friction Tubes

204. 10/06/1863 James Peeler of Bartow, GA for Method of Repairing Cotton Cards

205. 10/07/1863 William H. Hamilton of Mobile, AL for Candle Moulding Machine

206. 10/10/1863 E.S. Alexander of Moorfield, VA for Spur

207. 10/13/1863 George S. Smith of Keysville, VA for Loom

208. 10/19/1863 Henry Fitzgerald of Norfolk, VA for Fertilizer

209. 10/24/1863 W.N. (*W.H.) Smith of Richmond, VA for Cap Filling Machine

210. 10/26/1863 Isaac Mitchell of Columbus, GA for Cotton Seed Huller

211. 10/25/1863 (*10/28/1863) C. Williams of St. Louis, MO for Submarine Apparatus

212. 10/29/1863 B.H. Washington of Hannibal, MO for Bullet Swaging Machine

213. 10/31/1863 B.H. Washington of Hannibal, MO for Machine for Cutting and Serrating Friction Wire

214. 11/05/1863 B.H. Washington of Hannibal, MO for Machine for Finishing Cannon

215. 11/04/1863 E. Kempe, assignor to himself and Daniel Hagerty, of Richmond, VA for Grease Lamp

216. 11/18/1863 B.H. Washington of Hannibal, MO for Automatic Lathe

217. 11/27/1863 (*11/21/1863) Benjamin F. Holloway of Mobile, AL for Loom

218. 11/23/1863 B.H. Washington of Hannibal, MO for Machine for Casting Cannon

219. 11/24/1863 Louisiana Boykin of Glenville, AL for Mode of Preserving Meat

220. 12/24/1863 (*12/14/1863) J.P. Shaw, assignor to W.T. Chaffee, of Richmond, VA for Horse Shoe Machine

221. 12/29/1863 A.C. Jackson of Richmond, VA for Lubricating Compound

222. 01/02/1864 C.F. Richbourg of Gadsden, SC for Manufacture of Bagging

223. 01/06/1864 R.H.S. Thompson of New Orleans, LA for Projectile

224. 01/21/1864 Harry A. Bennet of Mecklenburg C.H., VA for Churn Dasher

225. 01/27/1864 John Scott of Ocala, FL for Truss

226. 01/27/1864 John Scott of Ocala, FL for Projectile

227. 02/05/1864 R.S. Sanxay and A. Gomert of Richmond, VA for Photographic Process for Duplicating Maps

228. 02/08/1864 Thomas C. Brown of Ballsville, VA for Sugar Cane Crusher

229. 02/10/1864 J.E. Garlington of Chambers C.H., AL for Water Boiler

230. 02/13/1864 John R. Maben of Allen Creek, VA for Machine for Cutting Shoe Pegs

231. 02/26/1864 C.E. Smart, Owings, and Taylor of Richmond, VA for Instruments for Fixing Sights to Cannon

232. 02/29/1864 George W. Powell of Clay Hill, AL for Breech Loading Cannon

233. 03/12/1864 J.E. Cargill of Columbus, GA for Ink Composition

234. 03/16/1864 John H. and J.C. Carlisle of Ringwood, NC for Instrument for Crooking Card Teeth

235. 03/29/1864 Y.A. Minnis of Franklinton, NC for Water Wheel

236. 03/30/1864 J.C. Patton and E. Cuthbert of Petersburg, VA for Torpedoes

237. 04/02/1864 M.L. Parry of Galveston, TX for Steam Condenser

238. 04/07/1864 David S. Cordle of Richmond, VA for Chain Link

239. 04/09/1864 J.E. Garlington of Chambers C.H., AL for Bridge

240. 04/15/1864 J.D. Gresit of Urbana, VA for Instrument for Calculating Distances

241. 04/18/1864 F.G. Smith of Columbia, TN for Sea-Going Vessel

242. 04/20/1864 B.H. Washington of Hannibal, MO for Machine for Making Wood Joints

243. 04/27/1864 J.B. Gayle of Larinburg, NC for Mode of Turning Eccentrics for Railroads

244. 05/06/1864 John H. and J.C. Carlisle of Ringwood, NC for Instrument for Renovating Card Teeth

245. 05/20/1864 A. Barbarin of New Orleans, LA for Torpedo

246. 06/31/1864 A.D. McCoy of Livingston, AL for Combined Bed and Tent

247. 01/04/1864 David Rawl of Lexington C.H., SC for Machine for Carding Cotton

248. 07/11/1864 William Moon of Richmond, VA for Clock Torpedo

249. 07/21/1864 James Price, assignor to J. and J.W. Tomkins of Tucker's Pond, SC for Tanning

250. 08/11/1864 G.W. and J.H. Wells of Charlottesville, VA for Artificial Leg

251. 07/12/1864 Anna Lewis, executrix of John Lewis, dec'd of Greensboro, NC for Machine for Ginning, Carding, and Spinning Cotton

252. 08/16/1864 Albert Strasser of Montgomery, AL for Artificial Leg

253. 08/25/1864 James D. Layton of Memphis, TN for Projectile

254. 08/29/1864 B.H. Washington of Hannibal, MO for Stethoscopic Probe

255. 09/06/1864 C. Williams of St. Louis, MO for Torpedo

256. 09/07/1864 C. Williams of St. Louis, MO for Torpedo

257. 10/05/1864 Richard M. Harvey of New Orleans, LA for Torpedo

258. 10/06/1864 C. Williams of St. Louis, MO for Submarine Boat

259. 10/25/1864 George W. Rains of Augusta, GA for Gunpowder Steaming Barrels

260. 10/26/1864 J.C. Patton of Petersburg, VA for Torpedo

261. 10/25/1864 C. Williams of St. Louis, MO for Submarine Boat

262. 11/10/1864 D.M. Somers and H. Dabney of Lynchburg, VA for Projectile

263. 11/16/1864 Thomas B. Taylor of Montgomery, AL for Machine for Making Cotton Rolls

264. 11/22/1864 A.T. Purejoy of Forrestville, NC for Wooden Shoe Sole

265. 11/23/1864 A.T. Purejoy of Forrestville, NC for Binding Last

266. 10/17/1864 W.N. Smith of Richmond, VA for Percussion Cap Rammer

267. 1/16/1865 William H. Wash of Lynchburg, VA for Machine for Making Percussion Caps

268. 1/17/1865 Joseph L. Seibert of Frederick, VA for Machine for Making Cartridge Cases for Spencer Rifles

269. 1/18/1865 Joseph L. Seibert of Frederick, VA for Machine for Flanging Spencer Rifle Cartridge Cases

270. 2/4/1865 James L. Gilbert & Charles W. Tracy of Sweetwater, AL for Wooden Shoe Soles

271. 2/18/1865 Samuel Pilsom of Danville, VA for Railroad Car Trunks

272. 3/4/1865 Charles A. McEvoy of Richmond, VA for Fuses

273. 3/14/1865 Nathaniel Walton of Cartersville, VA for Sugar Cane Crushers

274. 3/21/1865 Walter Husted Stevens of Richmond, VA for Pile Driver

RE1 10/28/1863 C. Williams of St. Louis, MO for Submarine Apparatus

 288

Examples of a Confederate Patent

John M. Brooke

Confederate States of America Patent No. 100
Filed May 2, 1862; issued June 29, 1862

No. 100
Confederate States of America

To all to whom these letters patent shall come:

Whereas John M. Brooke, of Richmond, Virginia, has alleged that he has invented a new and useful improvement in ships of war, which he states has not been known or used before his application; has made oath that he is a citizen of the Confederate States; that he does verily believe that he is the original and first inventor or discoverer of the said improvement, and that the same hath not, to the best of his knowledge and belief, been previously known or used; has paid into the treasury of the Confederate States the sum of forty dollars, and has presented a petition to the Commissioner of Patents, signifying a desire of obtaining an exclusive property in the said improvement, and praying that a patent may be granted for that purpose:

These are, therefore, to grant according to law to the said John M. Brooke, his heirs, administrators or assigns, for the term of fourteen years from the 29th of July, 1862, the full and exclusive right, and liberty of making, constructing, using, and vending to others to be used, the said improvement, a description whereof is given in the words of the said Brooke in the schedule hereunto annexed, and is made a part of these presents.

In testimony whereof, I have caused these letters to be made patent, and the seal of the Patent Office has been hereunto affixed.

Given under my hand at the city of Richmond, this 29th day of July, in the year of our Lord 1862

(signed) T.H. Watts
Attorney General

Countersigned and sealed with the seal of the Patent Office
Rufus R. Rhodes
Commissioner of Patents

 289

To all whom it may concern:

Be it known that I, John M. Brooke, a lieutenant in the Navy of the Confederate States, have invented a new and improved form of vessel, to be iron-clad, and if desired (armed) with cannon; and I do hereby declare that the following is a full and exact description thereof, reference being had to the annexed drawings making a part of this specification in which Figure I is a deck plan; Figure II a sheer plan, and Figure III a body plan.

The nature of my invention consists in so constructing the hull of the vessel that her bow and stern A and B, Figures I and II, shall each extend beyond the forward and after ends of the shield C, which protects crew and guns, sufficiently to give the sharpness necessary to the attainment of high speed, and the buoyancy to support the weight of iron covering the shield and sides of the vessel without increase of draft. Being submerged, all that part of the hull not covered by the shield is protected by the water from the projectiles of an enemy. The shield proposed for such improved form of vessel is of wood, covered on the exterior with iron, the surface inclined at such an angle as will permit the guns to be worked in the usual manner and yet deflect projectiles impinging upon it. This angle will be between 40 and 50. The eaves of the shield may be about two feet under water. To divide and prevent the water over the submerged part of the vessel from banking up at the forward or after ends of the shield in going ahead or astern, thereby retarding her progress and perhaps preventing the use of the bow or stern gun, a false bow and stern or tanks are constructed upon the submerged portion of the vessel corresponding more or less in form with the hull below. The false bow and stern may be decked, in which case they should not be so high above water as to interfere with firing of the bow and stern guns. These tanks may be used as reservoirs of water by which the draft of the vessel may be regulated at will. The stem, being submerged, may be fitted as a ram to strike the wooden bottoms of iron-clad vessels. This plan of construction is applicable in plating effectually ships built in the usual manner; it being simply necessary to remove the upper works and to cut them down forward and abaft the shield sufficiently to submerge the ends when down to the load-line, as illustrated in the case of the Confederate States steamer Virginia, which vessel was constructed in accordance with the plan herein set forth, furnished by me on the 23rd day of June, 1861, to the Honorable S.R. Mallory, Secretary of the Navy, to William P. Williamson, Chief-Engineer Confederate States Navy, and John L. Porter, Constructor Confederate States Navy, the two latter having been directed by the Honorable Secretary of the Navy, in conjunction with myself, to devise an iron-clad vessel. And this plan was applied to the Merrimac in preference to constructing a new vessel of eight or ten feet draft, in consequence of the impossibility of procuring in time boilers

 290

and engines suitable to the purpose. The boilers of the Merrimac were good, and as the chief-engineer was of opinion that the engines could be speedily repaired, it was considered expedient to apply the plan to her.

Claim

What I claim as my invention, and desire to secure by letters patent, consists in so constructing the hull of a vessel that her bow and stern shall each extend under water beyond the forward and after ends of the shield C, which protects the crew and guns, sufficiently to give the sharpness necessary to the attainment of high speed and the buoyancy to support the weight of iron applied without an inconvenient increase of draft.

John M. Brooke
Lieutenant C.S. Navy

Witnesses:
George Minor, Commander, C.S.N.
Charles J. Ost

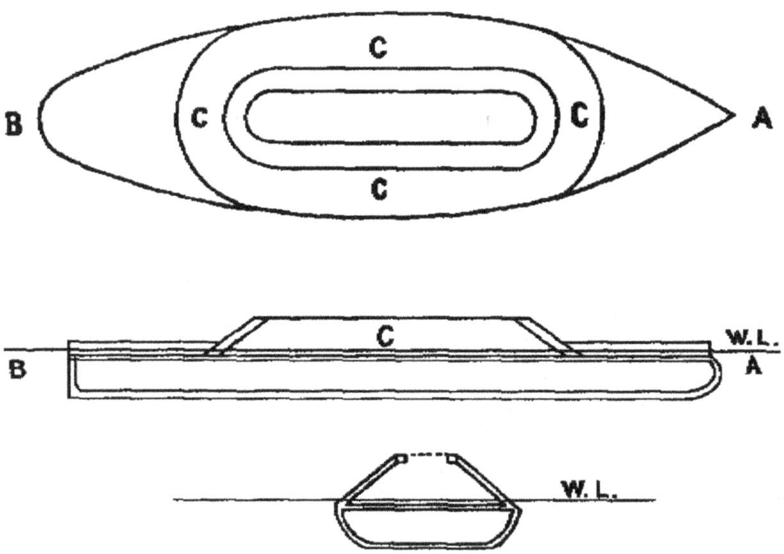

The three figures from the Brooke patent.

Example of a Civil-War Era United States Patent[3]

United States of America Patent No. 35,846 Issued July 8, 1862 No. 35,846
The United States of America

To All to Whom These Letters Patent Shall Come:

Whereas, Theodore R. Timby of Worcester, Massachusetts, has alleged that he has invented a new and useful Improvement in Revolving Battery-Towers, which he states has not been known or used before his application, has made oath that he is a Citizen of the United States, that he does verily believe that he is the original and first inventor or discoverer of the said Improvement and that the same hath not, to the best of his knowledge and belief been previously known or used, has paid into the treasury of the United States the sum of Thirty dollars and presented a petition to the Commissioner of Patents signifying a desire of obtaining an exclusive property in the said Improvement and praying that a patent may be granted for that purpose,

These are Therefore to grant according to law to the said Theodore R. Timby, his heirs, administrators or assigns for the term of seventeen years from the eighth day of July one thousand eight hundred and sixty two the full and exclusive right and liberty of making, constructing, using and vending to others to be used the said Improvement, a description whereof is given in the words of the said Theodore R. Timby in the schedule hereunto annexed and is made a part of these presents,

In Testimony whereof, I have caused these Letters to be made Patent and the Seal of the Patent Office has been hereunto affixed, Given under my hand at the City of Washington this eighth day of July in the year of our Lord one thousand eight hundred and sixty two and of the Independence of the United States of America the eighty seventh.

Caleb B. Smith
Secretary of the Interior

Countersigned and sealed with the Seal of the Patent Office
David P. Holloway
Commissioner of Patents

[3] The signed and sealed cover sheet of U.S. patents is only available on the original, as given to the inventor when the patent issues. This cover sheet has been reconstructed from other patents of the same period. There may be minor errors. For example, if the Secretary of the Interior or the Commissioner of Patents was absent from office on the day the patent issued, it would have been signed by an acting secretary or acting commissioner.

To all whom it may concern:

Be it known that I, Theodore R. Timby, of Worcester, in the county of Worcester and State of Massachusetts, have invented an Improvement in Revolving Towers, to be stationed on land or water for defensive or offensive warfare; and I do hereby declare that the following is a full, clear and exact description of the same, reference being had to the annexed drawings, making a part of this specification, in which

Figure 1 is an elevation of the tower, showing the embrasures and lookout; Fig. 2, a vertical section of the same; Fig. 3, a perspective view of the bed-plate and central shaft with its appendages; and Fig. 4, a top view of the commander's platform with its fixtures of telescope sight, hand-wheel, and signal-pulls.

In a revolving tower without special provisions the commander's platform or station must be fixed thereto and revolve therewith. When, therefore, it is desirable to concentrate the fire upon a fixed point or object, this necessitates, if that officer would keep his vision upon the object attacked, his constant progression in a direction contrary to that of the revolution of the tower. So, too, if the commander would keep a vigilant watch in all directions, as would be requisite in an attack upon water, or in an attack from both land and water, for instance, he would be compelled to fly from one point to another, and often, in consequence of the opposition revolution of the tower, his progress might be fatally retarded. It is my present purpose to overcome these difficulties, and with this view I combine with my revolving tower a central independently revolving shaft, upon the top of which, and in close proximity to the lookout, is arranged the commander's platform or station, which is under his constant control, and may at his option be revolved to the right or the left, or suffered to remain stationary, no matter what may be the condition of motion imparted to the tower, or whether it be in a state of rest.

To enable others skilled in the art to make and use my invention, I will proceed to describe its construction and operation.

My revolving tower, whether placed on land or water, is constructed entirely of iron or steel plates of any desirable thickness either in the particulars or aggregate; or it may be backed up or lined with timber, or with other suitable material. Its form may be cylindrical or conico-cylindrical, a truncated cone, or a pyramid, and it should be covered with a bombproof roofing, having a lookout at or near the top of the tower, which latter may consist of a slot or aperture through the walls or ramparts around its entire circumference, interrupted only at such points as is necessary to support the roof, which may be dome-formed, or something in the shape of a flattened bell, in which case, if preferred, the lookout may be formed through the barrel thereof. The tower

 293

may have one or more tiers of guns, but it will be found generally that one tier is all that a sea-going vessel will need or bear. The guns slide in and out of battery upon radial ways, and of course their extreme number in each tier will be regulated by the circumference of the walls. The tower revolves upon its foundation or base around its vertical axis by steam or other power, and through the instrumentality of well-known or other mechanical agents, so that all its guns may be concentrated upon any point of the compass, follow a moving object, or attack a stationary one, the time occupied in the revolution of the tower being sufficient to allow the guns to cool for repeated discharges. From within a hollow pivot which forms the eye or center of the foundation or base of the tower, and around which the tower revolves, rises a vertical shaft having proper bearings, whose top bears the commander's platform or station, and to which an independent rotation is given by means of a shaft with a hand-wheel attached to its upper end, extending down along the central shaft from the commander's station to a fixed collar attached to the said central shaft, where it is stepped, and has, a little above the step, a pinion which gears into a circular rack on the inside of the hollow pivot. This arrangement admits of an independent revolution of the platform at the will and by the exertion of the commander, and he can, of course, turn himself to any point desired while the tower continues its separate revolution or remains still.

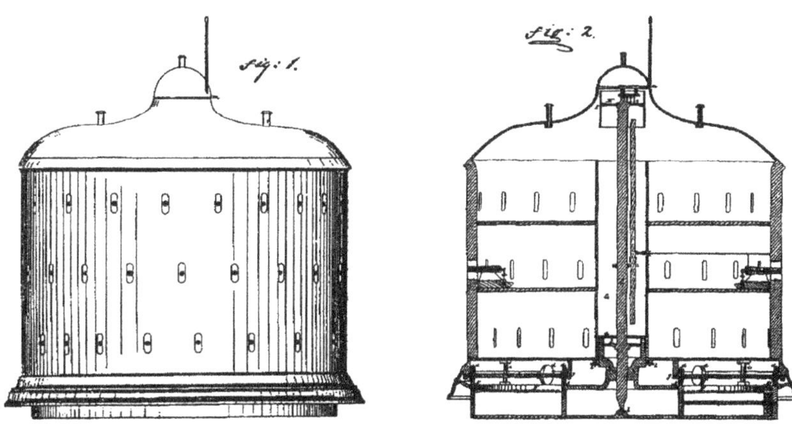

Figures 1 and 2 of Timby patent

Such is a general description of the improvement I now claim as new.

I shall next proceed to specify those elementary constituents which give to my combination its vitality, not meaning by this to limit myself to the exact mechanical means described for imparting independent action to the tower and central shaft, nor to confine myself to the precise arrangement represented, but to signify that, having pointed out and described one method of carrying into effect my invention, I shall claim all methods when the joint and individual operation of my tower and central shaft is as indicated.

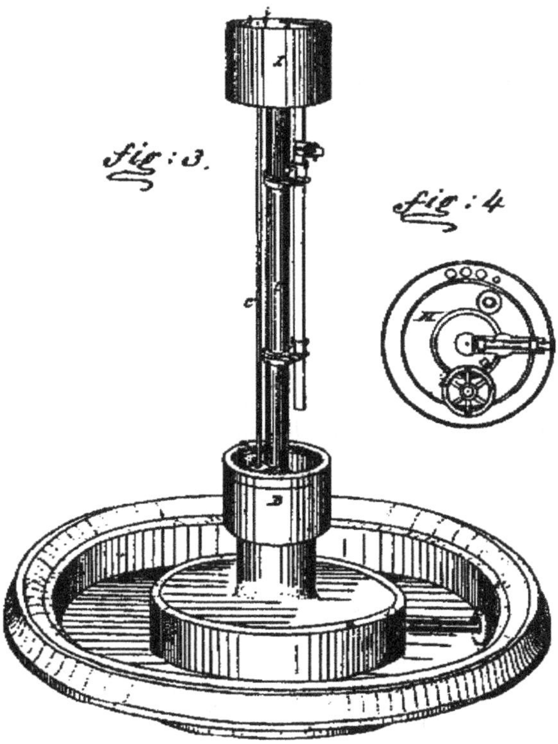

Figures 3 and 4 of Timby patent

In the case of a land-tower the ground is excavated for the purpose of receiving the strong masonry or iron foundation necessary to support the central railway-tracks A A', but for a water-tower the tracks, consisting of two concentric castings, cast whole or in sections, are located about three feet below the deck of the floating battery or gun-boat, so that only about two-thirds of the attitude of the tower will be exposed above. From the center of this foundation, or from the center of the track, rises the pivot or short shaft B, which is cast or otherwise made hollow to receive the circular rack a, pinion b, and shafts c and d. The driving-shaft D, carrying the pinion E, has its bearing in thc foundation upon which the railway-tracks are laid, and extends to the motor, which is situated from under the tower a proper distance, and on land is located in a bomb-proof vault. To the base of the tower is fastened the large wheel F, by means of the hangers c, having teeth on its under face with which the teeth of the pinion E mesh. The tracks A A' being of unequal radii, the friction rollers d revolve, each series independently of the other, on the shafts e, whose bearings are in the base of the revolving tower and in the hangers g. Friction-rollers l are also arranged around the well G, bearing against the pivot or short shaft B, to prevent lateral deflection of the tower as well as to lessen friction. The well or curb G, which is constructed concentrically with the axis of the tower, receives at its base the hollow pivot or shaft B, and through it and the hollow pivot or shaft passes the shaft C, which is stepped at h, and carries upon its upper end the platform H with its curb I, upon which the commander stands and turns himself to any desired point of the compass by means of the hand-wheel i, whose shaft C', extending along down side the shaft C, is stepped in the fixed collar or bearing k, and carries the pinion d, which meshes with the circular rack a. A lookout, m, is cut in the barrel of the dome above the commander's station or through any other point thereof, or through the top of the rampart or outer wall, the circumference of the platform H being increased and its elevation diminished accordingly.

From this description it will be observed that the commander of the tower may have an unobstructed view, the lookout m being practically unobstructed during the entire revolution of the tower, and that he may change the direction of his view at will or remain stationary, and this independently of the revolution of the tower.

Having thus described my invention and pointed out the manner in which it operates, what I claim therein as new, or desire to secure by Letters Patent of the United States, is –

A revolving tower for land or water, designed for offensive or defensive warfare, when combined with an independently rotating shaft, upon the top of which is arranged the commander's platform or station, as described. Theodore R. Timby

Witnesses:
John H. Shugert
Edm. F. Brown

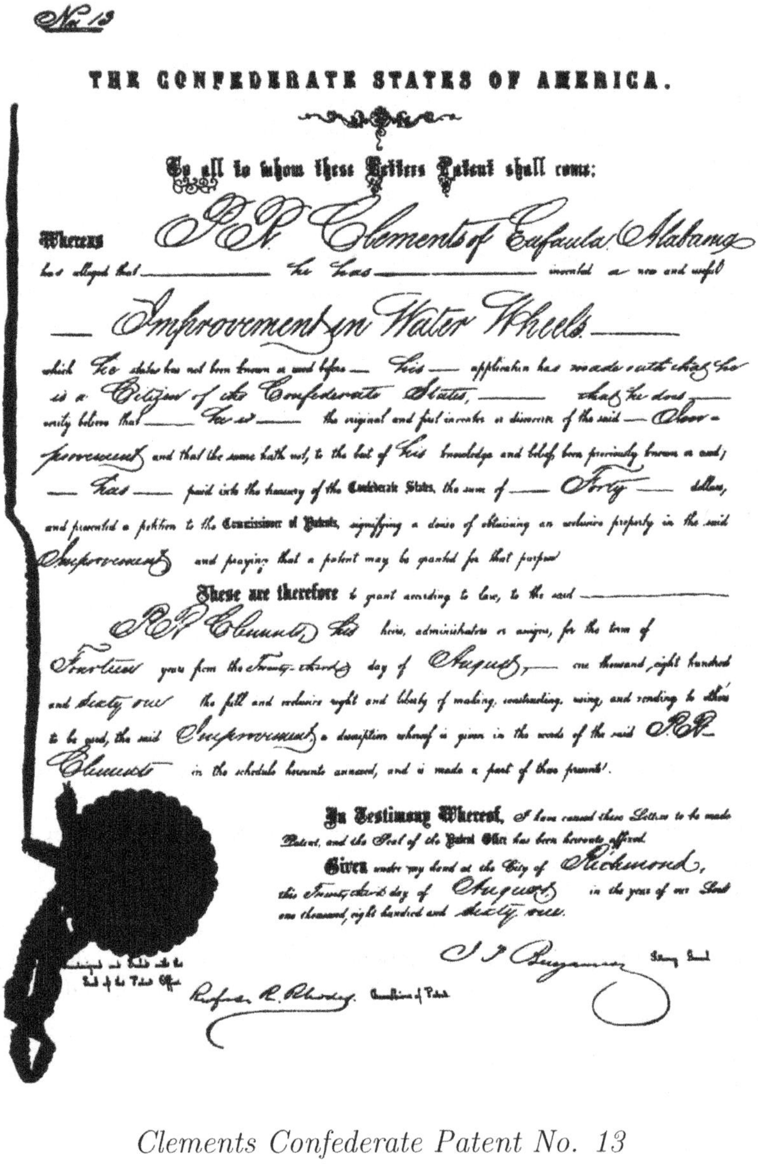
No. 13

THE CONFEDERATE STATES OF AMERICA.

To all to whom these Letters Patent shall come:

Whereas R. R. Clements of Eufaula Alabama has alleged that he has invented a new and useful Improvement in Water Wheels which he states has not been known or used before; he has made oath that he is a Citizen of the Confederate States, that he does verily believe that he is the original and first inventor or discoverer of the said Improvement and that the same hath not, to the best of his knowledge and belief been previously known or used; has paid into the treasury of the Confederate States, the sum of Forty dollars, and presented a petition to the Commissioner of Patents, signifying a desire of obtaining an exclusive property in the said Improvement and praying that a patent may be granted for that purpose.

These are therefore to grant according to law, to the said R. R. Clements his heirs, administrators or assigns, for the term of Fourteen years from the Twenty-third day of August, one thousand, eight hundred and Sixty one the full and exclusive right and liberty of making, constructing, using, and vending to others to be used, the said Improvement a description whereof is given in the words of the said R. R. Clements in the schedule hereunto annexed, and is made a part of these presents.

In Testimony Whereof, I have caused these Letters to be made Patent, and the Seal of the Patent Office has been hereunto affixed.

Given under my hand at the City of Richmond, this Twenty-third day of August in the year of our Lord one thousand, eight hundred and Sixty one.

J. P. Benjamin Attorney General

Countersigned and Sealed with the Seal of the Patent Office

Rufus R. Rhodes. Commissioner of Patents

Clements Confederate Patent No. 13

Finding Copies of Civil-War Era Patents

Although 274 patents were issued by the Confederate Patent Office, almost all of the records in the office were destroyed in the Evacuation Day fire near the end of the war. A copy of the specification and drawings for the individual patents was attached to the original patent document as issued to the inventor. Surprisingly few of these documents have been located, probably fewer than ten, although there may be many more in trunks in attics across the South. The general rule is, if you want a particular Confederate patent, you probably cannot find it.

Patents issued by the United States Patent Office (or Union Patent Office) during the war years (and before and after the war years) are much more accessible. Printed copies of almost all of the patents issued since 1836, and some issued before 1836, are available in the U.S. Patent and Trademark Office (PTO) Public Search Room in Alexandria, Virginia, in chronological order on microfilm, and also classified according to subject matter. These same patents are available, either in bound volumes or on microfilm, in Patent Depositories throughout the country.

Much easier in the age of the internet, digital files containing images of almost all (but not quite all) U.S. patents are available from websites, including from the U.S. Patent and Trademark Office website and from Google Patents, from which a searcher can print out copies. There are still a very few patents which do not fit into the system, and, unless the situation has recently changed, cannot be seen except on the PTO microfilms. Try teasing U.S. Patents 2,712,152 1/2 and 3,262,124 1/2 out of the digital files.

If one knows the patent number or date of issue of a patent being sought, a copy of the patent can be easily obtained. In fact, if the patent number is known, the PTO will currently supply a copy of it for $3.00 by mail. This price, which was once five cents during the 1930s, is still subject to change but has not changed recently.[4] In general, the only people who would buy and await delivery of a paper patent copy are those who cannot use the internet to get a free copy instantly.

There are easy printed indexes for twentieth century patents which allow one to find a patent knowing only the inventor's name or even the assignee's name. However, for patents from the Civil War era, and for some years afterwards, there are no easy printed indexes by inventor's name. The best that can be done in print media is to go to the Annual Report of the Commissioner of Patents year by year, where an annual index by inventor's name is

[4]See USPTO Fee Schedule, effective January 1, 2014, last revised January 1, 2016, published by the USPTO on the internet.

 298

given. A nearly complete set of these reports is available on open shelves in the Public Search Room in Alexandria. A complete set is available in the National Archives. Again, the internet comes to the rescue. Sites such as Google Patents allow searches, not only by inventor's name and issue date, but also a full search of the text of OCR translated printed patents. There are a few very old (pre-1836) patents around which were never set in type and printed, but which exist only as copies of handwritten documents. Computers cannot yet read those and translate them for the inquirer.

The original files for old issued patents are sometimes preserved, and arrangements can be made in the Attorney's Room at the Public Search Room in the PTO to see these files. There is much delay in obtaining old files, and a trip to the National Archives in Suitland, Maryland, may be necessary. In the modern digital PTO, the Patent Application Information Retrieval (PAIR) system maintains a digital record of the prosecution of the application, sometimes called the prosecution history or the file wrapper. There are over five million applications currently having PAIR files. When the information on the patent application becomes public, internet access to these records is available to the public through the PTO website or through Google Patents, as well as some other sites. These are records that may be used for private communication between the patent attorney and the patent examiner during prosecution of a patent application.

It is possible to deal with the PTO in the modern world without using the internet, but it is inconvenient, both for the applicant and for the PTO.

Bibliography

[Abb81] Henry L. Abbot. *The Beginning of Modern Submarine Warfare under Captain-Lieutenant David Bushnell.* Archon Books, Hamden, CT, 1881.

[Ada77] Charles Francis Adams, editor. *Memoirs of John Quincy Adams, comprising parts of his Diary from 1795 to 1848.* J.B. Lippincott, Philadelphia, PA, 1877.

[All87] Oliver E. Allen. The first family of inventors. *Invention and Technology*, 3(2), Fall 1987.

[Ame74] Mary Clemmer Ames. *Ten Years in Washington—Life and Scenes in the National Capital as a Woman Sees Them.* A.D. Worthington, Hartford, CT, 1874.

[Ano86] Anonymous. *Brief History of the United States Patent Office—1790 to 1886—With an Outline of Laws, Growth, Publications, Office Routine, etc.* R. Beresford, Printer, Washington, D.C., 1886.

[Arn59] Aza Arnold. The patent office and its mission. *The National Recorder of Religious and Literary Intelligence*, 2(1):29–3, February 1859.

[Ath54] Athenaeus. *Banquet of the Learned.* Henry G. Bohn, London, 1854.

[Aus82] Geoffrey D. Austrian. *Herman Hollerith, Forgotten Giant of Information Processing.* Columbia University Press, New York, NY, 1982.

[Ave82] James Hobson Aveling. *The Chamberlens and the Midwifery Forceps—Memorials on the Family and an Essay on the Invention of the Instruments (1882).* J. & A. Churchill, London, 1882.

[Bak17] Henry E. Baker. The Negro in the field of invention. *The Journal of Negro History*, 2:21–36, 1917.

[Bak76] Ronald J. Baker. *New and Improved ... Inventors and inventions that have changed the modern world.* British Museum Publications, London, 1976.

[Bar22] William Eleazer Barton. *The Life of Clara Barton, Founder of the American Red Cross.* Houghton Mifflin, New York, NY, 1922.

[BB43] Greville Bathe and Dorothy Bathe. *Jacob Perkins, His Inventions, His Times, His Inventions, His Times, & His Contemporaries.* The Historical Society of Pennsylvania, Philadelphia, PA, 1943.

[BB81] Eugenio Battisti and Filippo Brunelleschi. *Brunelleschi: The Complete Work.* Thames and Hudson, London, 1981.

[BDW88] Jacques-Pierre Brissot De Warville. *New Travels in the United States of America 1788.* Printed for J.S. Jordan, London, 1788.

[Bea28] John Davidson Beazley. *Greek Vases in Poland.* Clarendon, Oxford, 1928.

[Ben91] Park Benjamin. A United States patent. *Harper's Weekly*, 35, 4/11/1891 1891.

[Bet63] Royston Betts. *An Exposition of D.P. Holloway's Management of the Affairs of the Patent Office.* Author, Washington, D.C., 1863.

[BG90] Julian Parks Boyd and William H. Gaines. *The Papers of Thomas Jefferson.* Princeton University Press, Princeton, NJ, 1950–1990.

[Bla24] William Newnham Blane. *An Excursion through the United States and Canada During the Years 1822-23 By an English Gentleman.* Printed for Baldwin, Cradock, and Joy, London, 1824.

[BMF93] Lucy Etheldred Broadwood, J.A. Maitland, and J.A. Fuller. *English Country Songs, Words and Music.* Leadenhall Press, London, 1893.

[Boy35] Thomas Alexander Boyd. *Poor John Fitch—Inventor of the Steamboat.* G.P. Putnam's Sons, New York, NY, 1935.

[Boy60] John Boyle. Patents and civil rights in 1857–8. *Journal of the Patent Office Society*, 42:789–794, 1960.

[Bre26] J.A. Brearly. Old patent office models. *Journal of the Patent Office Society*, 8:280–85, 1926.

[Bro91] John M. Brooke. *The Virginia, or Merrimac: Her Real Projector.* W.E. Jones, Richmond, VA, 1891.

[Bro95] Noah Brooks. *Washington in Lincoln's Time.* Rinehart, New York, NY, 1895.

[Bro80] George M. Brooke. *John M. Brooke, Naval Scientist and Educator.* University Press of Virginia, Charlottesville, VA, 1980.

[Bry16] Wilhelmus B. Bryan. *History of the National Capital from its Foundation through the Period of Adoption of the Organic Act*, volume 2. Macmillan, New York, NY, 1914–1916.

[Buc90] Cassius M. Buck. *An Authentic Directory of the Model Room.* Author, Washington, D.C., 1890.

[But92a] Benjamin Butterworth. The effect of our patent system on the material development of the United States. In *Proceedings and addresses. Celebration of the beginning of the second century of the American patent system at Washington City, D.C. April 8, 9, 10, 1891*, pages 381–391, Washington, D.C., 1892. Press of Gedney & Roberts.

[But92b] Benjamin Butterworth, editor. *The Growth of Industrial Art.* Alfred A. Knopf, New York, NY, 1892.

[Cae39] Hans Paul Caemmerer. *A Manual on the Origin and Development of Washington, Government Printing Office.* Government Printing Office, Washington, D.C., 1939.

[Cam91a] Levin H. Campbell. *The Patent System of the United States so far as it relates to The Granting of Patents—A History.* McGill & Wallace, Washington, D.C., 1891.

[Cam91b] Levin H. Campbell. Power to grant patents for inventions—proceedings of the framers of the constitution in 1787. *Scientific American*, 64 ns(16):241, April 18 1891.

[Car45] George Calvin Carter. *Samuel Morey, The Edison of His Day.* Rumford Press, Concord, NH, 1945.

[CC69] Homer L. Calkin and Corrine Calkin. Iowa inventors and inventions. *The Palimpsest*, 50(7), July 1969.

[CH92] Henry Clay and James F. Hopkins. *Papers of Henry Clay.* University Press of Kentucky, Lexington, KY, 1959-1992.

[Cla15] Allen C. Clark. Dr. and Mrs. William Thornton. *Records of the Columbia Historical Society*, 18:144–208, 1915.

[Cla40] Allen C. Clark. Robert Mills, architect and engineer. *Records of the Columbia Historical Society*, 40-41:20ff., 1940.

[Co.94] Mercantile Illustrating Co. *Washington, D.C., with its Points of Interest Illustrated.* Mercantile Illustrating Co., New York, NY, 1894.

[Co.57] Bennett Machinery Co. *The Founder of Modern Industry.* Bennett Machinery Co., Clifton, NJ, 1957.

[Com92] Executive Committee. *Proceedings and addresses. Celebration of the beginning of the second century of the American patent system at Washington City, D.C. April 8, 9, 10, 1891.* Press of Gedney & Roberts, Washington, D.C., 1892.

[DC24] Cassius Dio Cocceianus. *Dio's Roman History.* Macmillan, New York, NY, 1924.

[DC63] L. Sprague De Camp. *The Ancient Engineers.* Doubleday, Garden City, NY, 1963.

[Del22] Judah Delano. *The Washington directory : showing the name, occupation, and residence, of each head of a family and person in business : the names of the members of Congress, and where they board : together with other useful information.* W. Duncan, Washington, D.C., 1822.

[Des71] John H. Dessauer. *My Years with Xerox.* Doubleday, Garden City, NY, 1971.

[Dil53] Leslie V. Dill. *The Obstetrical Forceps.* Thomas, Springfield, IL, 1953.

 303

[Dog22] Laurence L. Doggett. *History of the Young Men's Christian Association*, volume II. Association Press, New York, NY, 1922.

[DoSWE28] Bernard Duke of Saxe-Weimar Eisenach. *Travels through North America During the Years 1825 and 1826*. Carey, Lea & Carey, Philadelphia, PA, 1828.

[DotAR07] Connecticut Daughters of the American Revolution. *A memorial of the opening of the Ellsworth homestead at Windsor, Connecticut, October 8, 1903*. Tuttle, Morehouse & Taylor, New Haven, CT, 1907.

[Dun23] John M. Duncan. *Travels through part of the United States and Canada*. Hurst, Robinson, London, 1823.

[dW58] Raoul de Warren. *Les pairs de France sous l'ancien Régime*. Les cahiers nobles, Paris, 1958.

[Edi46] Editors. Fitch's steamboat. *Scientific American*, 2 os(4):1, October 17 1846.

[Edi48] Editors. The patent office. *Scientific American*, 3 os(36):285, May 27 1848.

[Edi51a] Editors. The patent office—its architect. *Scientific American*, 7 os(3):18, October 4 1851.

[Edi51b] Editors. The patent office—its architect. *Scientific American*, 6 os(49):387, August 23 1851.

[Edi51c] Editors. Patent office building. *Scientific American*, 7 os(5):38, October 18 1851.

[Edi53] Editors. Changes in the patent office. *American Polytechnic Journal*, 2:78, July 1853.

[Edi55a] Editors. Death of an inventor. *Scientific American*, 10 os(46):365, July 28 1855.

[Edi55b] Editors. History of the telegraph: Difficulties and success of an inventor. *Scientific American*, 11 os(3):19, September 29 1855.

[Edi57] Editors. State of affairs at the patent office. *Scientific American*, 13 os(16):125, December 26 1857.

[Edi58] Editors. Reminiscences of sewing machine inventors. *Scientific American*, 14 os(8):61, October 30 1858.

[Edi59a] Editors. The American patent office. *Scientific American*, 14 os(32):263, April 16 1859.

[Edi59b] Editors. Charles M. Keller and the American patent office. *Scientific American*, 14 os(37):310, May 21 1859.

[Edi60a] Editors. An inventor creating a sensation. *Scientific American*, 2 ns(5):67, January 28 1860.

[Edi60b] Editors. The patent office and its administration. *Scientific American*, 3 ns(22):345, November 24 1860.

[Edi60c] Editors. Photography at the patent office. *Scientific American*, 2 ns(5):73, January 28 1860.

[Edi60d] Editors. The United States patent office. *Scientific American*, 2 ns(25):386, June 16 1860.

[Edi61a] Editors. Letter from our Washington house. *Scientific American*, 4 ns(24):374, June 15 1861.

[Edi61b] Editors. Letter from Washington. *Scientific American*, 4 ns(22):342, June 1 1861.

[Edi61c] Editors. The "original" revolver—Colt's patent, etc. *Scientific American*, 4 ns(4):55, January 26 1861.

[Edi61d] Editors. Our editorial correspondence. *Scientific American*, 4 ns(17):266, April 27 1861.

[Edi61e] Editors. Patents in the seceded states. *Scientific American*, 4 ns(20):307, May 18 1861.

[Edi61f] Editors. Patents in the southern confederacy. *Scientific American*, 5 ns(2):19, July 13 1861.

[Edi61g] Editors. Patents in the southern confederacy. *Scientific American*, 5 ns(2):19, July 13 1861.

[Edi62] Editors. Business at the patent office. *Scientific American*, 7 ns(14):214, October 3 1862.

[Edi66a] Editors. Confederate states patent office. *Scientific American*, 14 ns(21):340, May 19 1866.

[Edi66b] Editors. Free and easy legislation on patents. *Scientific American*, 15 ns(2):23, July 7 1866.

[Edi67] Editors. A change at the patent office. *Scientific American*, 17 ns(26):407, December 28 1867.

[Edi68a] Editors. The board of examiners in the patent office. *Scientific American*, 19 ns(16):243, October 14 1868.

[Edi68b] Editors. Commissioner of patents' report. *Scientific American*, 18 ns(7):105, February 15 1868.

[Edi68c] Editors. Special correspondence of the scientific american— affairs at the patent office. *Scientific American*, 19 ns(12):180, September 16 1868.

[Edi69a] Editors. How contracts are awarded—gross frauds in the patent office. *Scientific American*, 20 ns(8):121, February 28 1869.

[Edi69b] Editors. The patent office before congress again. *Scientific American*, 20 ns(17):265, April 24 1869.

[Edi69c] Editors. Patent office contract swindle. *Scientific American*, 20 ns(10):152, March 6 1869.

[Edi70a] Editors. Bill to amend the patent laws now pending before congress. *Scientific American*, 22 ns(18):289, April 30 1870.

[Edi70b] Editors. Obituary—death of Thomas Ewbank. *Scientific American*, 23 ns(14):213, October 1 1870.

[Edi70c] Editors. Proposed plan for publishing patent office specifications and drawings. *Scientific American*, 22 ns(12):192, March 19 1870.

[Edi70d] Editors. Resignation of commissioner Fisher. *Scientific American*, 23 ns(21):328, November 19 1870.

[Edi70e] Editors. Speech of Hon. Thomas A. Jenckes, in defense of the patent system (delivered in the house, April 22, 1870). *Scientific American*, 22 ns(19):296–297, May 7 1870.

[Edi75] Editors. The Woodbury patent. *Scientific American*, 32 ns(2):16, January 1 1875.

[Edi78] Editors. Our Washington correspondence. *Scientific American*, 38 ns(11):164, March 16 1878.

[Edi80a] Editors. The past year's work in the patent office. *Scientific American*, 42 ns(5):65, January 31 1880.

[Edi80b] Editors. The patent bill now before the senate. *Scientific American*, 42 ns(12):176–77, March 20 1880.

[Edi80c] Editors. Two more patent bills. *Scientific American*, 42 ns(5):65, January 31 1880.

[Edi81] Editors. Resignation of the commissioner of patents. *Scientific American*, 45 ns(24):369, December 10 1881.

[Edi82] Editors. Nullification of patent laws. *Scientific American*, 46 ns(21):328, May 27 1882.

[Edi84] Editors. Patent office affairs. *Scientific American*, 50 ns(2):17, January 12 1884.

[Edi85] Editors. The beginning of the patent office. *Scientific American*, 53 ns(7):97, August 15 1885.

[Edi86] Editors. The Japanese assistant secretary of state. *Scientific American*, 54 ns(1):5, January 2 1886.

[Edi87] Editors. Celebration of the centennial of the enactment of the patent laws. *Scientific American*, 56 ns(11):160, March 12 1887.

[Edi91a] Editors. The first telegraph message and its author. *Electrical Engineer*, 12:201–2, August 19 1891.

[Edi91b] Editors. The patent centennial celebration—april 8, 9, and 10. *Scientific American*, 64 ns(16):243–45, April 18 1891.

[Edi97] Editors. Some curious old patents. *Scientific American*, 77 ns(24):370–71, April 11 1897.

[Edi15] Editors. Seventy years of the scientific american. *Scientific American*, 112:540 ff., June 5 1915.

[Edi23] Editors. [Charles Allen Munn, son of Orson Sr.]. *Journal of the Patent Office Society*, 6:460–61, 1923.

[Edi32] Editors. The new patent office quarters. *Journal of the Patent Office Society*, 14:289–301, April 1932.

[Edi80] Editors. Science and the citizen. *Scientific American*, 242(3):88–96, March 1980.

[Edi82] Editors. Science and the citizen. *Scientific American*, 246(5):92–101, May 1982.

[EE07] Oliver Evans and Thomas Ellicott. *The Young Mill-Wright and Miller's Guide*. Printed by Francis Baily, Octoraro, PA, 1807.

[Ell22] William Elliot. *The Washington guide : containing an account of the District of Colunbia ; the city of Washington ; abstract of the general laws of the corporation, alphabetically arranged ; the public buildings ; officers of the general government ; botany of the District of Columbia ; judiciary of the District*. S.A.Elliot, Washington, D.C., 1 edition, 1822.

[Ell26] William Elliot. *The Washington guide : containing an account of the District of Colunbia ; the city of Washington ; abstract of the general laws of the corporation, alphabetically arranged ; the public buildings ; officers of the general government ; botany of the District of Columbia ; judiciary of the District*. S.A.Elliot, Washington, D.C., 2 edition, 1826.

[Ell30] William Elliot. *The patentee's manual : containing a list of patents granted by the United States for the encouragement of arts & sciences, alphabetically arranged, from 1790 to 1830 : [to be continued by supplements] : also, the laws of Congress for granting patents : with a digest of all the decisions which have taken place in the courts of the United States, respecting patents*. S.A. Elliot, Washington, D.C., 1830.

[Ell37] William Elliot. *The Washington guide : containing an account of the District of Colunbia ; the city of Washington ; abstract of the general laws of the corporation, alphabetically arranged ; the public buildings ; officers of the general government ; botany of the District of Columbia ; judiciary of the District*. Franck Taylor. J. Crissy, Washington, D.C., 3 edition, 1837.

[Ell69] John B. Ellis. *The Sights and Secrets of the National Capital: A Work Descriptive of Washington City in all its Various Phases*. United States Publishing Company, New York, NY, 1869.

[E.M42] E.M.T. The patent office to leave Washington. *Journal of the Patent Office Society*, 24:66–69, January 1942.

[ES05] Oliver Evans and John Stevens. *The Abortion of the Young Steam Engineer's Guide.* Fry and Kammerer, Philadelphia, PA, 1805.

[Eva05] Oliver Evans. *The abortion of the young steam engineer's guide : containing an investigation of the principles, construction and powers of steam engines : a description of a steam engine on new principles . . . : a description of a machine, and its principles, for making ice and cooling water in large quantities . . . by the power of steam . . . : a description of four other patented inventions.* Printed for the author by Fry and Kammerer, Philadelphia, PA, 1805.

[Eva14a] Oliver Evans. *A Memorial, etc., to the Honorable, the Senators and Representatives in Congress.* s.n., Philadelphia, PA, 1814.

[Eva14b] Oliver Evans. *Patent right oppression exposed; or knavery detected.* The Booksellers, Philadelphia, PA, 1814.

[Eva17] Oliver Evans. *Oliver Evans to his counsel : who are engaged in defence of his patent rights, for the improvements he has invented : containing a short account of two out of eighty of his inventions, their rise and progress in despite of all opposition and difficulties : and two of his patents, with explanations.* Author, Philadelphia, PA, 1817.

[Eva19] George W. Evans. The birth and growth of the patent office. *Records of the Columbia Historical Society*, 22:104–124, 1919.

[Fau23] William Faux. *Memorable days in America : being a journal of a tour to the United States, principally undertaken to ascertain, by positive evidence, the condition and probable prospects of British emigrants, including accounts of Mr. Birkbeck's settlement in the Illinois, and intended to shew men and things as they are in America.* Printed for W. Simpkin and R. Marshall, London, 1823.

[FCM89] Benjamin Brown French, Donald B. Cole, and John J. McDonugh. *Witness to the Young Republic, 1828–1870; ed D.B. Cole & J.J. McDonough.* University Press of New England, Hanover, NH, 1989.

[Fea18] Henry Bradshaw Fearon. *Sketches of America—A Narrative of Journey of Five Thousand Miles through the Eastern and Western*

States of America. London, Printed for Longman, Hurst, Rees, Orme, and Brown, 1818.

[Fed32] P. J. Federico. The first patent act. *Journal of the Patent Office Society*, 14:237–252, April 1932.

[Fed36] P. J. Federico. Operation of the patent act of 1790. *Journal of the Patent Office Society*, 14:237–252, April 1936.

[Fed37] Bianca M. Federico. The patent office fire of 1836. *Journal of the Patent Office Society*, 19:804–833, 1937.

[Fen92] Robert W. Fenwick. The old and the new patent office. In *Proceedings and addresses. Celebration of the beginning of the second century of the American patent system at Washington City, D.C. April 8, 9, 10, 1891*, pages 453–471, Washington, D.C., 1892. Press of Gedney & Roberts.

[Fen29] Karl Fenning. The origin of the patent and copyright clause of the constitution. *Journal of the Patent Office Society*, 11:438–445, 1929.

[Fit76] John Fitch. *The Autobiography of John Fitch, ed. Frank D. Prager.* American Philosophical Society, Philadelphia, PA, 1976.

[Fle84] Tom Fletcher. *100 Years of the Negro in Show Business.* Da Capo Press, New York, NY, 1984.

[FN36] P. J. Federico and J. R. Nunn. A fragment of Texas history. *Journal of the Patent Office Society*, 18:407–410, 1936.

[For48] Peter Force. *American Archives for 1776*, volume 1 of 5. Matthew St. Clair Clarke, Washington, D.C., 1848.

[Fos28] B.G. Foster. Abraham Lincoln, the patent lawyer. *Journal of the Patent Office Society*, 10:458–465, 1928.

[Fow90] Francis Fowler. Memoir of Dr. Thomas P. Jones. *Journal of the Franklin Institute*, 10:1–7, July 1890.

[Fox47] Harold G. Fox. *Monopolies and Patents.* University of Toronto Press, Toronto, 1947.

[Fro77] Thomas Froncek. *An Illustrated History—The City of Washington.* Knopf, New York, NY, 1977.

[Fru43] M. Frumkin. The origin of patents. *Chamber's Journal*, pages 21–23, 1943.

[Ful55] Edmund Fuller. *Tinkers and Genius, The Story of the Yankee Inventors.* Hastings House, New York, NY, 1955.

[Gal34] Joseph Gales, Sr. *The Debates and Proceedings of the Congress of the United States.* Gales and Seaton, Washington, D.C., 1834.

[Gay40] Giovanni Gaye. *Carteggio inedito D'Artisti dei secoli XIV, XV, XVI publicato ed illustrato con documenti pure inediti.* Presso Giovani Molin, Florence, 1840.

[Gil78] Frank J. Gillis. *The Metamorphosis of a Derbyshire Ballad, Discourse in Ethnology,.* Indiana University Press, Bloomington, IN, 1978.

[Gle47] George Robert Gleig. *The Campaigns of the British Army at Washington and New Orleans in the Years 1814-1815.* J. Murray, London, 1847.

[Gol38] Harry Goldsmith. Abraham Lincoln, invention and patents. *Journal of the Patent Office Society*, 20:5–33, 1938.

[Gom46] Arthur Allen Gomme. *Patents of Inventions—Origin and Growth of the Patent System in Britain.* Published for the British council by Longmans Green and Co., London, 1946.

[Goo71] James M. Goode. *Capital Losses—A Cultural History of Washington's Destroyed Buildings.* Smithsonian Institution Press, Washington, D.C., 1971.

[Gre65] Frances Green. America's first patentee—Joseph Jenkes. *New England Galaxy*, 12(1):48–51, Summer 1965.

[Gri91] T.G. Gribble. Fossil patents. *Scientific American*, 65 ns(5):71–72, August 1 1891.

[GS49] Joseph Gales Sr. *Annals of the Congress of the United States—Second Congress.* Gales and Seaton, Washington, D.C., 1849.

[GW57] George C. Groce and David H. Wallace. *The New York Historical Society's Dictionary of Artists in America.* Yale University Press, New Haven, CT, 1957.

[Ham01] Margherita Arlina Hamm. *Famous families of New York, historical and biographical sketches of families which, in successive generations, have been identified with the development of the nation*, volume 2. G.P. Putnam's Sons, Remsen, 1901.

[Ham55] Talbot Hamlin. *Benjamin Henry Latrobe.* Oxford University Press, New York, NY, 1955.

[Ham77] Regina Combs Hammet. *History of St. Mary's County.* St. Mary's County Bicentennial Commission, Leonardtown, MD, 1977.

[Har92] Benjamin Harrison. Opening address by the President of the United States. In *Proceedings and addresses. Celebration of the beginning of the second century of the American patent system at Washington City, D.C. April 8, 9, 10, 1891*, page 23, Washington, D.C., 1892. Press of Gedney & Roberts.

[Har72] David Harbison. *Reaching for Freedom.* Scholastic Book Services, New York, NY, 1972.

[Har91] C.M. Harris. Specimens of genius and nicknacks—the early patent office and its museum. *Prologue*, 23:406–17, Winter 1991.

[HD94] P. Horsin-Deon. Norbert Rillieux. *The Louisiana Planter and Sugar Manufacturer*, pages 330–331, November 1894.

[Her81] Janet Sharp Hermann. *The Pursuit of a Dream.* Oxford University Press, Oxford, 1981.

[Hey76] Ernest V. Heyn. *Fire of Genius.* Anchor Press/Doubleday, Garden City, NY, 1976.

[Hog68] Charles Beecher Hogan. *The London Stage 1660-1800, Part 5*, volume 3. Southern Illinois Press, Carbondale, IL, 1968.

[How41] Humphry Howland. Address. *The Cultivator*, 8(12):199, December 1841.

[How44] Henry Howe. *Memoirs of the most eminent American mechanics : also, lives of distinguished European mechanics ; together with a collection of anecdotes, descriptions, &c. &c., relating to the mechanic arts.* Alexander V. Blake, New York, NY, 1844.

[How45] Henry Howe. *Historical collections of Virginia; containing a collection of the most interesting fact, traditions, biographical sketches, anecdotes, &c., relating to its history and antiquities, together with geographical and statistical descriptions. To which is appended an historical and descriptive sketch of the District of Columbia.* Babcock & Co, Charleston, SC, 1845.

[HP87] C.M. Harris and Daniel Preston. *Papers Relating to the Administration of the U.S. Patent Office During the Superintendency of William Thornton 1802-1828, A Guide to Accompany Federal Documentary Microfilm Edition No. 1.* National Archives, Washington, D.C., 1987.

[Hul95] E. Wyndham Hulme. The history of the patent system under the prerogative and at common law. *The Law Quarterly Review*, 46:141–54, April 1895.

[Hul97] E. Wyndham Hulme. On the consideration of the patent grant, past and present. *The Law Quarterly Review*, 51:313–18, July 1897.

[Hum78] David Hume. *The History of England from the Invasion of Julius Caesar to the Revolution in 1688.* Cadell and Davies, London, 1778.

[Hun64] T. Marshall Hunter. James Rumsey. *Virginia Cavalcade*, pages 33–40, Autumn 1964.

[Ing18] Charles A. Ingraham. Colonel Elmer E. Ellsworth, first hero of the civil war. *Wisconsin Magazine of History*, 1(4):349–374, June 1918.

[Ita83] Gen Itasaka. *Kodansha Encyclopedia of Japan.* Kodansha, New York, NY, 1983.

[Jam24] James I. *Statute 21*, chapter 3. s.n., London, May 24 1624.

[Jam89] Portia P. James. *The Real McCoy: African-American Invention and Innovation, 1619–1930.* Smithsonian Institution Press, Washington, D.C., 1989.

[Jap60] Sōryōjikan Japan. *The First Japanese Embassy—Japan-U.S. Centennial 1860–1960.* Consulate General of Japan, New York, NY, 1960.

[Jef40] Eber Jeffery. Nothing left to invent. *Journal of the Patent Office Society*, 22:479–481, 1940.

[JEJ34] Charles Nicolas Jenks, Esek Esten, and John Whipple Potter Jenks. *History and Trees of the Jenks Family*. Author, Boston, MA, 1934.

[Jen65] Amy La Follette Jensen. *The White House and its Thirty-Four Families*. McGraw Hill, New York, NY, 1965.

[[Jo39] [Joseph Cloud]. Horse hay rake. *The Farmer's Cabinet*, 3:226, February 1839.

[Joh60] James D. Johnston. *China and Japan: Being a Narrative of the Cruise of the U.S. Steam-Frigate Powhatan in the Years 1857, '58, '59 and '60, Including an Account of the Japanese Embassy to the United States*. C. Desilver; Baltimore, Cushings & Bailey, Philadelphia, PA, 1860.

[Jon28] Thomas P. Jones. Editorial note. *Journal of the Franklin Institute*, 1 (ns):414, June 1828.

[Jon71] Stacy V. Jones. *The Patent Office*. Praeger, New York, NY, 1971.

[Jor11] John W. Jordan. Sellers family. In John W. Jordan, editor, *Colonial and Revolutionary Families of Philadelphia*, volume 2, New York, NY, 1911. The Lewis Publishing Company.

[Kal46] Clarence A. Kalk. Remaining units of the patent office to return to Washington. *Journal of the Patent Office Society*, 28:459, July 1946.

[Kei74] De B. Randolph Keim. *Keim's Illustrated Guide to the Museum of Models*. De B. Randolph Keim, Washington, D.C., 1874.

[Ken53] George W. Kendall. *A sketch of the history of the Ohio Mechanic's Institute : and a statement of its present condition*. A. Pugh, Cincinnati, OH, 1853.

[Ket37] Charles Frances Kettering. *Centennial Celebration of the American Patent System 1836–1936*. Government Printing Office, Washington, D.C., 1937. On behalf of The National Committee on the Centennial Celebration of the American Patent System.

[Kna30] Samuel Lorenzo Knapp. *Sketches of Public Characters Drawn from the Living and the Dead With Notices of Other Matters.* E. Bliss, New York, NY, 1830.

[Kni79] George H. Knight. *Address and memorial in opposition to the bill (S. no. 300 and H.R. no. 1612) "to amend the statutes relating to patents and for other purposes" : read before and adopted by the Cincinnati Board of Trade, December 18th, 1878.* Times Book and Job Printing Establishment, Cincinnati, OH, 1879.

[Kni91a] George H. Knight. Letter to the editor: The father of the patent office. *Scientific American*, 64 ns(19):295–96, May 9 1891.

[Kni91b] George H. Knight. *The Patent Franchise in the U. S.* Temple Court, New York, NY, 1891.

[Kni11] H. Jackson Knight. *Confederate Invention.* Louisiana State University Press, Baton Rouge, LA, 2011.

[Kur59] Harry Kursh. *Inside the U.S. Patent Office.* Norton,, New York, NY, 1959.

[Kyl02] Joanna Nicholls Kyle. The humorous side of the patent office. *Overland Monthly*, 39:890–96, May 1902.

[Lad02] Story B. Ladd. Patents in relation to manufactures. In *Census Report, Twelfth Census of the United States*, volume 10, pages 749–978. United States Bureau of the Census, Washington, D.C., 1902.

[Lan87] Charles Lanman. *Biographic Annals of the Civil Government of the United States.* J. Anglim, Washington, D.C., 2 edition, 1887.

[Lat72] George A. Latimer. *A sketch of the life of Oliver Evans.* J.C. Harkness, Wilmington, DE, 1872.

[Lay87] Edwin T. Layton. The most original (James Rumsey). *Invention and Technology*, pages 50–56, Spring 1987.

[Lip03] Andrew A. Lipscomb. *The Writings of Thomas Jefferson.* Thomas Jefferson Memorial Association, Washington, D.C., 1903.

[Log02] Mary Simmerson Logan. *Thirty years in Washington, or, Life and scenes in our national capital : Portraying the wonderful operations in all the great departments, and describing every important function of our national government ... With sketches of the*

315

presidents and their wives ... from Washington's to Roosevelt's administration. A.D. Worthington, Hartford, CT, 1902.

[LW82] Rayford W. Logan and Michael R. Winston. *Dictionary of American Negro Biography.* Norton, New York, NY, 1982.

[Mab43] Carleton Mabee. *The American Leonardo, A Life of Samuel F.B. Morse.* A.A. Knopf, New York, NY, 1943.

[Mac56] Thomas Babington Macaulay. *History of England.* Brampton Society, London, 1856.

[Mag80] W.A. Maguire. Major General Ross and the burning of Washington. *The Irish Sword*, 14(55):117–128, Winter 1980.

[Mar79] Bayly Ellen Marks. *Economics and Society in a Staple Plantation System, St. Marys County.* PhD thesis, University of Maryland, 1979.

[Mas70] Charles Mason. *What shall be done with the Surplus Funds of the Patent Office?* Government Printing Office, Washington, D.C., 1870.

[McC67] M.J. McCosker. *The Historical Collection of Insurance Co. of North America.* Beck Engraving Co., Philadelphia , PA, 1967.

[McM15] John Bach McMaster. *A history of the people of the United States from the Revolution to the Civil War.* D. Appleton, New York, NY, 1915.

[Mer68] James E. Merna. U.S. patent office finds a new home—crosses the Potomac to Crystal City, Virginia. *Journal of the Patent Office Society*, 50:191–200, March 1968.

[Mil39] Kelly Miller. The Negro Stephen Foster. *The Etude*, page 431ff., July 1939.

[Mit92] Charles Elliot Mitchell. Birth and growth of the American patent system. In *Proceedings and addresses. Celebration of the beginning of the second century of the American patent system at Washington City, D.C. April 8, 9, 10, 1891*, pages 43–55, Washington, D.C., 1892. Press of Gedney & Roberts.

[ML89] Sir Henry C. Maxwell-Lyte. *A History of Eton College 1440–1884.* Macmillan and Co., London, 1889.

[ML11] Sir Henry C. Maxwell-Lyte. *A History of Eton College 1440–1910.* Macmillan and Co., London, 1911.

[Moo55] J.G. Moore. *Patent Office and Patent Laws, or, a Guide to Inventors and a Book of Reference for Judges, Lawyers, Magistrates and Others.* Parry & M'Millan, Philadelphia, PA, 1855.

[Mor00] Edward L. Morse. The District of Columbia's part in the early history of the telegraph. *Records of the Columbia Historical Society*, 3:161–179, Winter 1900.

[Mot57] Frank Luther Mott. *Scientific American*, pages 318–324. Harvard University Press, Cambridge, MA, 2 edition, 1957.

[MSD83] Roy M. Marcot, Charles R. Suydam, and Gerald Denning. *Spencer Repeating Firearms.* Northwood Heritage Press, Irvine, CA, 1983.

[MT12] Edward B. Moore and Frederick A. Tennant. *History of the Patent System, Hearings before House Committee on Patents.* Government Printing Office, Washington, D.C., 1912.

[Mun60] Munn & Co. *Patent laws of the United States ; together with rules and proceedings in the Patent Office ; also hints to inventors, how to obtain letters patent in the United States and Europe ; with a synopsis of the patent of foreign countries, and advise relating thereto, also useful legal information to inventors and patentees.* Munn & Co., New York, NY, 1860.

[Mun24] Munn & Co. *Patents and Trademarks.* Munn & Co., New York, NY, 1924.

[Mur54] Norimasa Muragaki. The diary of Muragaki Awaji-no-kami. *Contemporary Japan*, 22,23:679ff., 147ff., 1954.

[Nee84] Samuel H. Needles. The governor's mill, and the globe mills. *Pennsylvania Magazine of History and Biography*, page 279ff., 1884.

[New30] Jane Elizabeth Newton. A forgotten chapter of Confederate history. *Journal of the Patent Office Society*, 12:248–258, 1930.

[Nor87] Joseph J. Norris. *Some Important Patents, Trademarks, and Innovations of Blacks.* Author, Arlington, VA, 1987.

[Off77] United States Patent Office. *An Authentic Account of the Fire of September 24, 1877 which Destroyed the North and West Halls of the United States Patent Office Building; Also, some Historical Data in Reference to the Organization and Early History of the Patent Office from 1790 to 1877.* Government Printing Office, Washington, D.C., 1877.

[Off41] United States Patent Office. *United States Patent Law Sesquicentennial Celebration—A Record of the Proceedings Commemorating the One Hundred and Fiftieth Anniversary of the Signing of the First United States Patent Law.* Government Printing Office, Washington, D.C., 1941.

[otCoP29] Annual Report of the Commissioner of Patents. *Annual Report of the Commissioner of Patents.* Government Printing Office, Washington, D.C., 1829.

[otCoP30] Annual Report of the Commissioner of Patents. *Annual Report of the Commissioner of Patents.* Government Printing Office, Washington, D.C., 1830.

[otCoP33] Annual Report of the Commissioner of Patents. *Annual Report of the Commissioner of Patents.* Government Printing Office, Washington, D.C., 1833.

[otCoP37] Annual Report of the Commissioner of Patents. *Annual Report of the Commissioner of Patents.* Government Printing Office, Washington, D.C., 1837.

[otCoP38] Annual Report of the Commissioner of Patents. *Annual Report of the Commissioner of Patents.* Government Printing Office, Washington, D.C., 1838.

[otCoP41] Annual Report of the Commissioner of Patents. *Annual Report of the Commissioner of Patents.* Government Printing Office, Washington, D.C., 1841.

[otCoP49] Annual Report of the Commissioner of Patents. *Annual Report of the Commissioner of Patents.* Government Printing Office, Washington, D.C., 1849.

[otCoP50] Annual Report of the Commissioner of Patents. *Annual Report of the Commissioner of Patents.* Government Printing Office, Washington, D.C., 1850.

[otCoP58] Annual Report of the Commissioner of Patents. *Annual Report of the Commissioner of Patents.* Government Printing Office, Washington, D.C., 1858.

[otCoP69] Annual Report of the Commissioner of Patents. *Annual Report of the Commissioner of Patents.* Government Printing Office, Washington, D.C., 1869.

[otCoP78] Annual Report of the Commissioner of Patents. *Annual Report of the Commissioner of Patents.* Government Printing Office, Washington, D.C., 1878.

[otCoP79] Annual Report of the Commissioner of Patents. *Annual Report of the Commissioner of Patents.* Government Printing Office, Washington, D.C., 1879.

[otCoP87] Annual Report of the Commissioner of Patents. *Annual Report of the Commissioner of Patents.* Government Printing Office, Washington, D.C., 1887.

[otCoP88] Annual Report of the Commissioner of Patents. *Annual Report of the Commissioner of Patents.* Government Printing Office, Washington, D.C., 1888.

[otCoP90] Annual Report of the Commissioner of Patents. *Annual Report of the Commissioner of Patents.* Government Printing Office, Washington, D.C., 1890.

[Par68] James Parton. *People's Book of Biography; or Short Lives of the Most Interesting Persons of All Ages and Countries.* A.S. Hale, Hartford, CT, 1868.

[Par57] C.N. Parkinson. *Parkinson's Law (and Other Studies in Administration).* Houghton Mifflin Co., Boston, 1957.

[Pay90] Henry M. Paynter. The first U.S. patent. *Invention and Technology*, pages 18–22, Fall 1990.

[Pet69] Gaius Petronius. *Satyricon.* Harvard University Press, Cambridge, MA, 1969.

[Pla92] Orville H. Platt. Invention and advancement. In *Proceedings and addresses. Celebration of the beginning of the second century of the American patent system at Washington City, D.C. April 8, 9,*

10, 1891, pages 57–76, Washington, D.C., 1892. Press of Gedney & Roberts.

[Plu23] William Plumer. *William Plumer's Memorandum of Proceedings in the United States Senate, 1803-1807; ed. E.S. Brown.* MacMillan, New York, NY, 1923.

[Poe61] Jessie J. Poesch. *Titian Ramsay Peale, 1799-1885, and His Journals of the Wilkes Expedition.* American Philosophical Society, Philadelphia, PA, 1961.

[Poo86] Benjamin Perley Poore. *Perley's Reminiscences of Sixty Years in the National Metropolis.* Hubbard Bros., Philadelphia, PA, 1886.

[Pos73] Robert Charles Post. *Patents, Physics and Politics: The Washington Career of Charles Grafton Post 1838-1868.* PhD thesis, UCLA, 1973.

[Pos76] Robert Charles Post. *Patents, Physics and Politics: The Washington Career of Charles Grafton Post.* Science History Publications, New York, NY, 1976.

[Pow78] Howard Benjamin Powell. *Philadelphia's First Fuel Crisis.* Pennsylvania State University Press, University Park, PA, 1978.

[Pra46] Frank D. Prager. Brunelleschi's patent. *Journal of the Patent Office Society*, 28:109ff., 1946.

[Pra55] Frank D. Prager. The steamboat pioneers before the founding fathers. *Journal of the Patent Office Society*, 37:486–522, 1955.

[Pra58] Frank D. Prager. The steam boat interference 1787–1793. *Journal of the Patent Office Society*, 40:611–643, 1958.

[Pre85] Daniel Preston. The administration and reform of the U.S. patent office 1790-1836. *Journal of the Early Republic*, 5:331–353, Fall 1985.

[Pri75] Samuel Irenæus Prime. *The Life of Samuel F.B. Morse, LL.D., Inventor of the Electro-Magnetic Recording Telegraph.* D. Appleton, New York, NY, 1875.

[Pri06] William Hyde Price. *The English Patents of Monopoly.* Harvard University Press, Cambridge, MA, 1906.

[Pro31] John Clagett Proctor. Stories of early lower ninth street ... patent office site. *The Sunday Star*, page 9 ff., 1931.

[Pro32] John Clagett Proctor. *The Fenwick Family*, volume 5, pages 1117–1122. Lewis Historical Pub. Co, New York, NY, 1932.

[Pro35] John Clagett Proctor. Old patent office Parthenon—part of Burnes tract, etc. *The Sunday Star*, pages F–2, 1935.

[Pro39] John Clagett Proctor. Old patent office structure associated with growth of government, etc. *The Sunday Star*, December 10 1939.

[Pro42] John Clagett Proctor. Early homes of the U.S. patent office. *The Sunday Star*, February 1 1942.

[PS62] Caius Plinius Secundus. *The natural history of Pliny*, volume 6. Henry G. Bohn, London, 1862.

[PS70] Frank D. Prager and Gustina Scaglia. *Brunelleschi: Studies of His Technology and Inventions.* MIT Press, Cambridge, MA, 1970.

[Ran60] Ruth Painter Randall. *Colonel Elmer Ellsworth.* Little, Brown, Boston, MA, 1960.

[Ray88] William Chandler Raymond. *Curiosities of the Patent Office.* W.C. Raymond, Syracuse, NY, 1888.

[Rea70] David Read. *Nathan Read : his invention of the multi-tubular boiler and portable high-pressure engine, and discovery of the true mode of applying steam-power to navigation and railways ; a contribution to the early history of the steamboat and locomotive engine.* Hurd and Houghton, New York, NY, 1870.

[Rei71] Daniel D. Reiff. *Washington Architecture 1791–1861. Problems in Development.* U.S. Commission of Fine Arts, Washington, D.C., 1971.

[Rem81] Robert V. Remini. *Andrew Jackson and the Course of American Freedom.* Harper & Row, New York, NY, 1981.

[Rho61] Rufus R. Rhodes. *Rules and Directions for Proceedings in the Confederate States Patent Office.* Mitchie and Dunnavant, Richmond, VA, 1861.

[Rho63] Rufus Rhodes. *Annual Report of the Commissioner of Patents.* Ritchie & Dunnavant, Printers, Richmond, VA, 1863.

[Rho65] Rufus R. Rhodes. *Four Annual Reports of the Confederate Patent Office (1862–1865).* CSA Congress, Richmond, VA, 1862–1865.

[Ril51] Henry Hiram Riley. A scene at the patent office. *The Knickerbocker or New York Monthly Magazine*, 37:13–18, January 1851.

[Ros28] Joseph Rossman. Abraham Lincoln. *Journal of the Patent Office Society*, 10:292–296, 1928.

[Rum88] James Rumsey. *A short treatise on the application of steam : whereby is clearly shewn, from actual experiments, that steam may be applied to propel boats or vessels of any burthen against rapid currents with great velocity ; great velocity [sic] ; the same principles are also introduced with effect, by a machine of a simple and chep [sic] construction, for the purpose of raising water sufficient for the working of grist-mills, saw-mills, &c. and for watering meadows and other purposes of agriculture.* J. James, Philadelphia, PA, 1788.

[Sas89] Samuel Sass. A patently false patent myth. *The Skeptical Inquirer*, 10:310–12, 1989.

[Sch43a] Jean Schneider. The patent business in 1791. *Journal of the Patent Office Society*, 25:603–4, August 1943.

[Sch43b] Cathleen Schurr. Two hundred years of patents and copyrights. *American History Illustrated*, 50:60–71, August 1943.

[SD01] Ezra Stiles and Franklin Bowditch Dexter. *The Literary Diary of Ezra Stiles*, volume 3. C. Scribner, New York, NY, 1901.

[Sel84] George Escol Sellers. Early engineering reminiscences. *American Machinist*, pages 3–4, July 12 1884.

[Sel65] George Escol Sellers. *Early Engineering Reminiscences (1815–40) of George Escol Sellers, ed Eugene S. Ferguson.* Smithsonian Institution Press, Washington, D.C., 1965.

[Sel69] Charles Coleman Sellers. *Charles Willson Peale.* Scribner, New York, NY, 1969.

[Sel80] Charles Coleman Sellers. *Mr. Peale's Museum.* Norton, New York, NY, 1980.

[She82] Vivian M. Sherlock. *The Fever Man, a Biography of Dr. John Gorrie.* Medallion Press, Tallahassee, FL, 1982.

[Sie92] Donna H Siemiatkoski. *The ancestors and descendants of Chief Justice Oliver Ellsworth and his wife, Abigail Wolcott : and the story of Elmwood, their homestead.* Gateway Press, Baltimore, MD, 1992.

[Sin07] Ester Singleton. *The Story of the White House.* McClure, New York, NY, 1907.

[Sin74] Bruce Sinclair. *Philadelphia's Philosopher Mechanics—A History of the Franklin Institute 1824–1865.* Johns Hopkins University Press, Baltimore, MD, 1974.

[SJM+20] C. Shibama, James D. Johnston, Shigehiko Miyoshi, Nichi-Bei Kyōai, and Muragaki Awaji-no Kami. *The First Japanese Embassy to the United States of America.* America-Japan Society, Tokyo, 1920.

[Ski46] Ichabod Lord Skinner. *American Journal of the Improvements in the Useful Arts and Mirror of the Patent Office.* William Greer, Washington, D.C., 1846.

[Slu87] Patricia Carter Sluby. *Creativity and Inventions.* Research Unlimited, Arlington, VA, 1987.

[Smi32] Harriet Smither, editor. *Journals of the fourth Congress of the Republic of Texas, 1839-1840, to which are added the relief laws.*, volume 3. Von Boeckmann-Jones, Austin, TX, 1932.

[Smi59] Arthur M. Smith. A. Lincoln, inventor. *Journal of the Patent Office Society*, 41:447–457, 1959.

[Smi77] Merritt Roe Smith. *Harpers Ferry Armory and the New Technology, The Challenge of Change.* Cornell University Press, Ithaca, NY, 1977.

[Spr80] Mary Mace Spradling. *In Black and white : a guide to magazine articles, newspaper articles, and books concerning more than 15,000 Black individuals and groups.* Gale Research Co., Detroit, MI, 3rd, supplement edition, 1980.

[SS43] William E. Smith and Ophia D. Smith. *Continued Study Units in Cultural Life: Colonial Inventions*, volume 1. F.A. Davis, Philadelphia, PA, 1943.

[SSL94] George Smith, Leslie Stephen, and Sidney Lee, Sir. *The Dictionary of national biography.* Oxford University Press, London, 1894.

[Sta01] United States Department Of State. *History of the Department of State of the United States.* Government Printing Office, Washington, D.C., 1901.

[Sta12] Navy League Of The United States. *John Fitch—The first in the World's History to Invent and Apply Steam Propulsion of Vessels through water.* R.S. Peck, Hartford, CT, 1912.

[Sta68] Stan Stanton. The patent office of the confederacy. *The Arlington Historical Magazine*, pages 53–60, October 1968.

[SW63] Floyd Stovall and Walt Whitman, editors. *Walt Whitman—Prose Works 1892*, volume 1. New York University Press, New York, NY, 1963.

[Swa56] W.A. Swanberg. *Sickles the Incredible.* Scribner, New York, NY, 1956.

[SY76] Elinor Stearns and David N. Yerkes. *William Thornton, a Renaissance Man in the Federal City.* American Institute of Architects Foundation, Washington, D.C., 1976.

[Tag24] Samuel Taggart. Letters of Samuel Taggart, representative in congress, 1803–1814. In *Proceedings of the American Antiquarian Society*, volume 33, pages 297–438, Worcester, MA, 1924. American Antiquarian Society.

[Tak36] Korekiyo Takahashi. *Takahashi Korekiyo jiden [Autobiography of Takahashi Korekiyo].* Chūō Kōron Shinsha, Tokyo, 1936.

[Tho14] William Thornton. *Short account of the origin of steam boats : written in 1810, and now committed to the press.* Printed by Rapine and Elliot,, Washington, D.C., 1814.

[Tol71] Frederick B. Tolles. *Sybilla Masters: Notable American Women*, volume II. Harvard University Press, Cambridge, MA, 1971.

[Tom01] D.A. Tompkins. *The Cotton Gin—The History of its Invention.* The Author, Charlotte, NC, 1901.

[Top71] Edgar A. Toppin. *A Biographical History of Blacks in America Since 1528.* David McKay, New York, NY, 1971.

[TS32] Frances Milton Trollope and Donald Arthur Smalley. *Domestic Manners of the Americans.* A.A. Knopf, New York, NY, 1832.

[Tuc21] Max W. Tucker. The patent office of the confederacy. *Journal of the Patent Office Society*, 3:596–600, 1921.

[TW73] Mark Twain and Charles Dudley Warner. *The Gilded Age: a tale of to-day.* American Publishing Co., Hartford, CT, 1873.

[Twa89] Mark Twain. *A Connecticut Yankee in King Arthur's Court.* Collier, New York, NY, 1889.

[Var57] John Varden. *A Guide for Visitors to the National Gallery.* Commissioner of Patents by Pelkinhorn's Steam Printing Plant, Washington, D.C., 1857.

[Ver92] Laurie Verge. All roads lead to Surrat house—unfortunately. *The Maryland Line*, 13:3, November 1992.

[Wal38] Thomas Ustick Walter. *Report on the new Treasury buildings and Patent Office at Washington made at the request of the Committee of the House of Representatives on Public Buildings and Grounds.* L.R. Bailey, Philadelphia, PA, 1838.

[Web24] Gustavus A. Weber. *The Patent Office, Its History, Activities and Organization.* Johns Hopkins Press, Baltimore, MD, 1924.

[Weg63] Frederick Wegner. *Submarine Fighter of the American Revolution—The Story of David Bushnell.* Dodd, Mead, New York, NY, 1963.

[WF88] James Grant Wilson and John Fiske. *Appleton's Cyclopedia of American Biography*, volume 5, page 255. D. Appleton and Company, New York, NY, 1888.

[Wha03] Anne Hollingsworth Wharton. *Social Life in the Early Republic.* J.B. Lippincott, Philadelphia, PA, 1903.

[Whi84] James Terry White, editor. *The National Cyclopedia of American Biography.* James T. White, Clifton, NJ, 1984.

[Wil57] John S. Williams. *History of the Invasion and Capture of Washington.* Harper & Bros, New York, NY, 1857.

[WME61] Walt Whitman and Edwin Haviland Miller (Ed.). *The Correspondence, 1842–1867.* New York University Press, New York, NY, 1961.

[Wri92] Carroll D. Wright. The relation of invention to labor. In *Proceedings and addresses. Celebration of the beginning of the second century of the American patent system at Washington City, D.C. April 8, 9, 10, 1891*, pages 77–110, Washington, D.C., 1892. Press of Gedney & Roberts.

[Wro26] Lawrence C. Wroth. *Abel Buell of Connecticut: Silversmith, Type Founder & Engraver.* Yale University Press, New Haven, CT, 1926.

[WS63] Walt Whitman and Floyd Stovall. *Prose Works 1892.* New York University Press, New York, NY, 1963.

[Wym15] William I. Wyman. The patent office and invention since 1845: How the government has kept pace with the inventor. *Scientific American*, 112:533, June 5 1915.

[Wyn50] James Wynne. *Lives of Eminent Literary and Scientific Men of America.* D. Appleton, New York, NY, 1850.

[Yan84] Dorothy Cowser Yancey. The Stuart double plow and double scraper: The invention of a slave. *The Journal of Negro History*, 69(1):48–52, 1984.

Index

Made in the USA
Monee, IL
16 September 2026

63296746R00193